中国北方草地植物
彩色图谱(续编)

Atlas of Rangeland Plants in Northern China
(Supplement)

谷安琳　　王宗礼　主　编
Editors in Chief　Gu Anlin & Wang Zongli

中国农业科学技术出版社
China Agricultural Science and Technology Press

图书在版编目（CIP）数据

中国北方草地植物彩色图谱：续编/谷安琳，王宗礼主编.—北京：中国农业科学技术出版社，2011.6
ISBN 978-7-5116-0488-0

Ⅰ.①中…Ⅱ.①谷…②王…Ⅲ.①草地-植物-中国-图谱
Ⅳ.①Q948.52-64

中国版本图书馆CIP数据核字(2011)第100693号

责任编辑　黄　卫　贺可香
责任校对　贾晓红
出版发行　中国农业科学技术出版社
　　　　　北京市中关村南大街12号　　邮编：100081
电　　话　(010)82109704(发行部)；(010)82109711(总编室)
网　　址　http://www.castp.cn
经 销 商　新华书店北京发行所
印 刷 者　北京东方宝隆印刷有限公司
开　　本　880mm×1230mm　1/32
印　　张　11.75
印　　数　1~2700册
字　　数　350千字
版　　次　2011年6月第1版　2011年6月第1次印刷
定　　价　120元

版权所有●翻印必究
本书摄影作品版权归原作者所有，未注明作者的图片均为谷安琳拍摄
本书所有图片未经许可，不得引用

《中国北方草地植物彩色图谱（续编）》

主　　编	谷安琳　王宗礼
副 主 编	吴新宏　王育青　赵利清
编　　委	朱进忠　卢欣石　温刘君　辛有俊　易　津
	余晓光　李卫军　周青平　张洪江　李　鹏
英文翻译	谷安琳　温刘君
英文审校	Jack Carlson
摄　　影	谷安琳　赵利清　赵　凡　卢欣石　张洪江
	拾　涛　吴新宏　易　津　乔　江

《Atlas of Rangeland Plants in Northern China (Supplement)》

Editors in Chief
　　Gu Anlin　Wang Zongli
Vice-Editors in Chief
　　Wu Xinhong　Wang Yuqing　Zhao Liqing
Assistant Editors
　　Zhu Jinzhong　Lu Xinshi　Wen Liujun　Xin Youjun　Yi Jin
　　Yu Xiaoguang　Li Weijun　Zhou Qingping　Zhang Hongjiang
　　Li Peng
English Translators
　　Gu Anlin　Wen Liujun
English Reviser
　　Jack Carlson
Photograph Contributors
　　Gu Anlin　Zhao Liqing　Zhao Fan　Lu Xinshi
　　Zhang Hongjiang　Shi Tao　Wu Xinhong　Yi Jin　Qiao Jiang

目录
Contents

前言
Preface ··· I

使用说明
Guide to the use of this book ·· III

正文
Text of plant atlas ··· 1~350

附录: 中国地理区域示意图
Appendix: Geographical regions of China ······································ 351

参考文献
References ·· 352

中文名索引
Index of Chinese Names ·· 354

拉丁名索引
Index of Latin Names ··· 356

拉丁名及中文名分科索引
Index of Latin and Chinese Names in Families ······························· 360

前言

　　《中国北方草地植物彩色图谱》已经出版一年多了，得到许多同仁和读者的首肯与赏识。

　　由于在《中国北方草地植物彩色图谱》中的许诺，《中国北方草地植物彩色图谱(续编)》通过作者努力，又与大家见面了。本书共收录了中国北方草地较为常见和地区特有的野生植物350种(含亚种和变种)，隶属于56科201属。两本合计收录了68科346属850种(含亚种和变种)。

　　本书仍然沿用前书的版式与体裁风格。每种植物描述包括中文名、拉丁名，以及简要的形态特征、花期、地理分布、生境和用途；英文对照除英文名外其他为对应翻译。本书每种植物都有1~3张野外实地拍摄的照片。

　　中国草地孕育着藻类、真菌、苔藓、地衣、蕨类、裸子和被子植物约15 000种，仅北方草地常见的亦近2 000种。以上两本只拾缀不足千种，且更多地集中在北方中东部的草原与荒漠草地中，挂一漏万在所难免，西北与青藏草地植物图谱期待再次补编出版。感谢许多同仁提出按分类、按区域排序描述等建议，也待统一再版时加以考虑。

　　本书仍然得到了国家自然科技资源平台项目"牧草种质资源标准化整理、整合及共享试点"子项目和农业部牧草种质资源收集项目"雀麦属牧草种质资源考察与收集保护"课题的资助。植物学前辈，武素功先生、刘书润先生、陈山先生和王六英先生，以及地方草地科学工作者和美国合作专家继续给予了帮助与支持。谨向关怀、支持本书出版的所有个人和单位表示衷心感谢。

　　期待本书能再次得到大家的认可和青睐，也期待读者的批评指正。

编　者

2010年11月

Preface

The "Atlas of Rangeland Plants in Northern China" has been in print for more than a year and we are grateful it has been well received by those in our profession and others with interest in the plants of our grasslands.

With our commitment expressed in the first volume to develop the atlas to the fullest feasible extent, we have compiled a second volume, "Atlas of Rangeland Plants in Northern China (Supplement)". It describes and illustrates an additional 350 wild plant (species, subspecies or varieties) in 201 genera and 56 families, including common and locally endemic species. Together, the two volumes include 850 plants (species, subspecies and varieties) in 346 genera and 68 families.

The format and style of the sequel are the same as the first. The description of each plant includes the Chinese name, scientific name, brief morphology, flowering time, geographic distribution, habitat and practical uses, with one to three color photos taken of the plant in the field. The English section has been translated from the Chinese, except for the English common names.

About 15 000 plant species naturally occur on the rangelands of China, including algae, epiphytes, bryophytes, lichens, ferns, gymnosperms, and angiosperms. Even common, widespread species in the northern rangelands number more than 2 000. The 850 plants of the two volume atlas are mostly distributed in the steppe and desert communities of the central-east areas of the region. Therefore, additional volumes have been proposed and are expected to represent the northern west extending through Qinghai and Tibet. We appreciate the advice and suggestions from experts on the organization of the atlas taxonomically and by regions, which we intend to put to good use in the upcoming editions.

As with the first volume, this volume has been financially sponsored by the national project for Forage Germplasm Resources, Standardization, Systematization, and Information Sharing, and also by the Ministry of Agriculture project Brome Germplasm, Investigation, Collection, and Protection. The atlas continues to involve many contributing experts: botanists of the elder generation, professors Wu Sugong, Liu Shurun, Chen Shan, and Wang Liuying, specialists from local and regional rangeland stations, and cooperating American scientists. We are grateful to all individuals and organizations supporting the development and publishing of this atlas.

We trust the second volume of the atlas will be as well received as the first. As always we encourage reader comments and suggestions for continued improvement.

Editors
November, 2010

使用说明

1. 书中给出了每种植物的中文名、拉丁名和所属科名。名称主要依据《中国植物志》中文版，部分依据地方植物志，禾本科小麦族植物名称依据中国植物志禾本科英文版《Flora of China》Vol. 22。

2. 为了方便国外读者，书中给出了植物的英文名或拉、汉（或其他语言）名称的英译。对在英语国家已有英文名的植物，其英文名引自英文原著和相关文献。对没有英文名的植物，则根据其拉丁名、中文名或其他语言名称意译为英文。

3. 植物以花色排序，同种植物可能有不同的花色，以最常见的花色为主。有的植物没有明显的花冠或典型的花被片，则以其花序中最明显的特征颜色排序。本书中收录的木贼科、柏科、麻黄科、荨麻科、藜科、大戟科、车前科、水麦冬科和禾本科植物均放在了绿色部分中，莎草科植物放在了褐色部分中。

4. 同花色植物中，科以植物分类学恩格勒系统排序，属、种则按拉丁名字母顺序排列。

5. 本书在述及该植物的分布地域时，凡属我国境内，均置于分号之前，分号之后属境外。

6. 植物在国内的分布使用省和自治区名称描述，分布范围比较广时，用地理区域描述，如"华中"、"西北"。书中的地理区域不同于行政大区，有的省区可能跨不同的地理区域（附录）。

7. 书末附有植物中文名索引和拉丁名索引；同时还附有以恩格勒系统排序的分科索引，同科中的属、种按拉丁名字母顺序排列。

Guide to the use of this book

1. Each species in this book is listed by its Chinese name, Latin name and family name, which are mostly from the Chinese edition the Flora of China and partly from local floras. Names of species in Triticeae are according to English edition of the Flora of China, Vol. 22. Poaceae.

2. For the convenience of foreign readers, the species common name or the English translation of the Latin or Chinese (or other language) name is listed. For species with a common name in English-speaking countries, the name is from English-written floras or related references. Species without an English common name are given a translation of the Latin scientific name or the Chinese name (or another language's name).

3. Species are arranged by flower color. Some species contain populations with different flower colors, but they are arranged under the color which is of the most common presentation. For species lacking a showy corolla or typical perianth, they are arranged under the color which is of the character color of their inflorescences. As so, the species in families of Horsetail, Cypress, Ephedra, Nettle, Goosefoot, Spurge, Plantain, Arrowgrass and Grass included are found in the green section; and the species in family of Sedge in the brown section.

4. Within the same flower color group, families are arranged according to the Engler system of plant taxonomy; genera and species within each family are arranged in alphabetical order.

5. In the description of species distribution, regions (areas) within the territory of China are arranged before semicolon; other countries and regions (areas) outside the territory of China are arranged after semicolon.

6. Country, province and autonomous region names are used for conveying distributions of the plants; larger distribution ranges are defined by geographical regions, such as "C China" (Central China), "NW China" (Northwest China), which are different than administration regions. Some provinces and autonomous regions are in different geographical regions (Appendix).

7. Index of Chinese names and Index of Latin names are listed at the end of the atlas. Index of Latin and Chinese names arranged in families according to the Engler system also is listed, in alphabetical order by Latin names of genus, and species.

问荆
Equisetum arvense L.
木贼科
Equisetaceae

【特征】多年生草本；茎直立，二型；孢子茎高5～40cm，无叶绿素，肉质，不分枝，孢子散后枯萎；营养茎高20～60cm，绿色，轮生分枝向上，总宽不超过10cm；孢子囊穗圆柱状，长1.5～4cm。
【分布】东北、华北、西北、华中、西南；北半球温带。
【生境】草甸、林地灌丛、湖边和溪旁湿润地。
【用途】药用。对牲畜有毒。

Field horsetail
Horsetail family
Perennial herb; stems erect, dimorphic; fertile stems 5～40 cm tall, achlorophyllus, fleshy, unbranched, withering after shedding spores; sterile stems 20～60 cm tall, green, branching in whorls, sharply ascending, less than 10 cm wide overall; cones cylindric, 1.5～4 cm long.
Distribution: NE, N, NW, C and SW China; temperate zone in the Northern Hemisphere
Habitat: Meadows, woodland thickets, moist sites along lakeshores and stream banks
Use: Medicine. Poisonous to animals

林问荆 (林木贼)
Equisetum sylvaticum L.
木贼科
Equisetaceae

【特征】多年生草本；茎直立，二型；孢子茎高20～30cm，初时无叶绿素，肉质，不分枝，后为绿色，有分枝，孢子散后仍存活；营养茎高30～70cm，绿色，轮生分枝向两侧伸展，总宽达30cm，侧枝再数次分枝；孢子囊穗圆柱状，长0.8～2.5cm。

【分布】黑龙江、吉林、山东、内蒙古、新疆；日本、欧洲、北美洲。

【生境】阴湿溪边、泥沼林地。

【用途】药用。

Woodland horsetail
Horsetail family
Perennial herb; stems erect, dimorphic; fertile stems 20～30 cm tall, at first achlorophyllus, fleshy and unbranched, later green and branched, persisting after cone falling; sterile stems 30～70 cm tall, green, branching in whorls, laterally spreading to 30 cm wide overall, lateral branches further branching several times; cones cylindric, 0.8～2.5 cm long.
Distribution: Heilongjiang, Jilin, Shandong, Inner Mongolia and Xinjiang; Japan, Europe, North America
Habitat: Shady wet stream banks and boggy woods
Use: Medicine

杜松
Juniperus rigida Sieb. et Zucc.
柏科
Cupressaceae

【特征】小乔木或灌木，高达11m；树冠塔形或圆柱形；叶质厚，挺直，条状刺形，3叶轮生，基部有关节，上面凹下，白粉带位于凹槽中，下面具脊，横断面呈"V"状；球花单生于叶腋；球果球形，被白粉。花期5月。

【分布】东北、华北；朝鲜、日本。

【生境】海拔1 400～2 200m的山坡、坡顶、岩缝。

【用途】家具、农具用材；庭院绿化；药用。

Needle juniper
Cypress family
Small tree or shrub to 11 m tall; crown pyramidal or columnar; leaves thick and stiff, linear-spinelike, in whorls of 3, jointed at base, excavated and with white-powdery groove above, ribbed beneath, transverse cut V-shaped; cone solitary in axils; strobilus globose, glaucous. Flowering May.
Distribution: NE and N China; Korea, Japan
Habitat: Mountain slopes, hilltops and rock crevices at 1 400～2 200 m
Use: Materials for furniture and farm tools; courtyard planting; medicine

侧柏
Platycladus orientalis (L.) Franco
柏科
Cupressaceae

【特征】乔木,高达20m;幼树树冠卵状尖塔形,老树树冠圆形;鳞叶;球果卵圆形,成熟前肉质,蓝绿色,被白粉,成熟后木质,开裂,红褐色。花期5月。
【分布】东北、华北、华东、华中、华南和西南大部分省区;朝鲜。
【生境】海拔250~3 300m的山地阳坡、岩缝。
【用途】建筑、家具、农具用材;庭院绿化;药用。

Oriental arborvitae
Cypress family
Tree to 20 m tall; crown ovoid-pyramidal when young, rounded when old; leaves scale-like; cones ovoid, fleshy and glaucous when immature, woody, dehiscent and reddish-brown when ripe. Flowering May.
Distribution: Most regions in NE, N, E, C, S and SW China; Korea
Habitat: Mountain sunny slopes and rock crevices at 250~3 300 m
Use: Materials for buildings, furniture and farm tools; courtyard planting; medicine

拍摄人:赵利清 Photos by Zhao Liqing

拍摄人：拾 涛 Photo by Shi Tao

新疆方枝柏
Sabina pseudosabina
(Fisch. et C. A. Mey.) W. C. Cheng et W. T. Wang.
(*Juniperus pseudosabina* Fisch. et C.A. Mey)
柏科
Cupressaceae

【特征】灌木，高达3～4m；枝平铺或斜升，小枝直或弧曲，方圆形或四棱形；具鳞叶和刺叶；球果卵圆形，成熟时褐黑色或蓝黑色，多少有白粉，种子1粒。花期5～6月。
【分布】新疆；蒙古、西伯利亚、中亚。
【生境】海拔1 500～3 000m的林缘、灌丛、石质山坡。
【用途】水土保持；庭院绿化。

Xinjiang juniper
Cypress family
Shrub to 3～4 m tall; branches procumbent or ascending, branchlets straight or arcuate, quadrate-rounded or quadrangular; leaves both scale-like and needle-like; cones ovoid, brown-black or blue-black when ripe, somewhat glaucous, with 1 seed. Flowering May to June.
Distribution: Xinjiang; Mongolia, Siberia, Central Asia
Habitat: Forest margins, thickets and rocky slopes at 1 500～3 000 m
Use: Soil conservation; courtyard planting

叉子圆柏
Sabina vulgaris Antoine
(沙地柏 *Juniperus sabina* L.)
柏科
Cupressaceae

【特征】匍匐灌木,高不足1m,稀小乔木;枝密,斜上伸展;具鳞叶和刺叶;球果倒三角状球形或叉状球形,成熟前蓝绿色,被白粉,成熟后褐色、紫蓝色或黑色,多少有白粉。花期5月。
【分布】西北,内蒙古西部;蒙古、俄罗斯(西伯利亚、远东)、中亚、南欧。
【生境】海拔1 100~2 800m的多石山坡、针叶林或混交林、沙丘。
【用途】固沙和水土保持;药用。

Savin
Cypress family
Procumbent shrub, less than 1 m tall, rare small tree; branchlets dense, ascending; leaves both scale-like and needle-like; cones obdeltoid-globose or forked-globose, glaucous when immature, brown, purple-blue or black and somewhat glaucous when ripe. Flowering May.
Distribution: NW China, W Inner Mongolia; Mongolia, Russia (Siberia and Far East), Central Asia, S Europe
Habitat: Stony slopes, coniferous and mixed forests, and dunes at 1 100~2 800 m
Use: Fixing dunes and soil conservation; medicine

拍摄人:赵利清 Photos by Zhao Liqing

拍摄人：赵利清 Photos by Zhao Liqing

木贼麻黄
Ephedra equisetina **Bunge**
麻黄科
Ephedraceae

【特征】灌木，高1~1.5m；木质茎直立；小枝稠密，蓝绿色或灰绿色，具细槽纹；叶膜质鞘状，2浅裂，基部增厚呈红色；雌球花通常2枚对生节上，成熟时苞片肉质，红色；珠被管稍弯曲；种子通常1粒。花期6~7月。
【分布】华北、西北；蒙古、西伯利亚、中亚、高加索。
【生境】干旱区的山脊、山顶、岩壁。
【用途】药用。

Mongolian ephedra (Bluestem joint fir)
Ephedra family
Shrub 1~1.5 m tall; woody stems erect; branchlets dense, glaucous, finely sulcate; leaves membranous, sheath-like, 2-lobed, basally thickened and reddish; ovulate cones usually 2 per node, bracts fleshy and red at maturity; integument tube slightly curved; seed usually 1 per cone. Flowering June to July.
Distribution: N and NW China; Mongolia, Siberia, Central Asia, Caucasia
Habitat: Hill ridges, hilltops and cliffs in arid areas
Use: Medicine

单子麻黄
Ephedra monosperma C. A. Mey
麻黄科
Ephedraceae

【特征】草本状矮小灌木，高3～15cm；木质茎短小，多分枝；当年小枝绿色，开展，常弯曲，具浅沟纹；叶膜质鞘状，2裂，裂片略增厚；雌球花单生或对生节上，成熟时苞片肉质，红色；珠被管长而弯曲；种子1粒。花期6月。
【分布】华北、西北，黑龙江、四川、西藏；蒙古、俄罗斯（西伯利亚、远东）。
【生境】石质山坡、干燥沙地。
【用途】药用。

One-seed ephedra
Ephedra family
Dwarf and herbaceous shrub, 3～15 cm tall; woody stems short, much branched; new branchlets green, spreading, usually curved, with shallow grooves; leaves membranous, sheath-like, 2-lobed, lobes slightly thickened; ovulate cones 1 or 2 per node, bracts fleshy and red at maturity; integument tube long and curved; seed 1 per cone. Flowering June.
Distribution: N and NW China, Heilongjiang, Sichuan and Tibet; Mongolia, Russia (Siberia and Far East)
Habitat: Rocky slopes and dry sands
Use: Medicine

拍摄人：赵利清 Photo by Zhao Liqing

拍摄人：赵利清 Photo by Zhao Liqing

斑子麻黄
Ephedra rhytidosperma Pachom.
麻黄科
Ephedraceae

【特征】垫状灌木，高10～20cm；木质茎灰褐色，枝节膨大，坚硬，小枝绿色，密集在节上，呈辐射状排列，具粗槽纹；叶膜质鞘状，2裂；雌球花单生，成熟时苞片肉质，红色；种子2粒。花期5～6月。

【分布】内蒙古（贺兰山）、宁夏、甘肃；蒙古。

【生境】荒漠地带的石质山坡和坡麓。

【用途】药用。

Scalyseed ephedra
Ephedra family

Cushion-like shrub 10～20 cm tall; woody stems grey-brown, branch nodes swollen and hard; branchlets green, aggregated on nodes, radially arranged, roughly sulcate; leaves membranous, sheath-like, 2-lobed; ovulate cones solitary, bracts fleshy and red at maturity; seeds 2 per cone. Flowering May to June.

Distribution: Inner Mongolia (Helan Mountains), Ningxia and Gansu; Mongolia
Habitat: Rocky slopes and foothills in desert zone
Use: Medicine

狭叶荨麻
Urtica angustifolia Fisch. ex Hornem.
荨麻科
Urticaceae

【特征】多年生草本，高40～150cm，全株被短柔毛和螫毛；茎直立，单一或稍分枝，四棱形；单叶对生，披针形，边缘具齿；穗状或狭圆锥状花序腋生；花被片4裂。花期7～8月。
【分布】东北、华北；蒙古、朝鲜、日本、俄罗斯（西伯利亚、远东）。
【生境】山地、沙丘、湿地、林缘、灌丛、溪旁。
【用途】饲用；药用；纤维材料；造纸；提炼栲胶；嫩叶可食。

Narrowleaf nettle
Nettle family
Perennial herb 40～150 cm tall, pubescent and stinging-hairy throughout; stems erect, single or slightly branched, quadrangular; simple leaves opposite, lanceolate, serrate; spikes or narrow panicles axillary; perianth 4-parted; Flowering July to August.
Distribution: NE and N China; Mongolia, Korea, Japan, Russia (Siberia and Far East)
Habitat: Mountains, dunes, wetlands, forest edges, thickets and streamsides
Use: Forage; medicine; fiber materials; papermaking; extracting tannin; young leaves edible

麻叶荨麻
Urtica cannabina L.
荨麻科
Urticaceae

【特征】多年生草本,高100~200cm,全株被柔毛和螫毛;茎直立,丛生,具纵棱;单叶对生,掌状3深裂或全裂,裂片羽状深裂或羽状缺刻;穗状聚伞花序腋生;花被片4裂。花期7~8月。
【分布】东北、华北、西北、四川;蒙古、俄罗斯、欧洲。
【生境】山坡、沟壑、路旁、居民点附近。
【用途】饲用;药用;嫩叶可食。

Hempleaf nettle
Nettle family
Perennial herb 100~200 cm tall, pubescent and stinging-hairy throughout; stems erect, tufted, longitudinally angulate; simple leaves opposite, palmately 3-parted or divided, segments pinnately parted or incised; spicate cymes axillary; perianth 4-parted; Flowering July to August.
Distribution: NE, N and NW China, and Sichuan; Mongolia, Russia, Europe
Habitat: Slopes, ravines, roadsides and residential places
Use: Forage; medicine; young leaves edible

拍摄人：赵利清 Photos by Zhao Liqing

华北大黄
***Rheum franzenbachii* Münter**
蓼科
Polygonaceae

【特征】多年生草本，高30～90cm；茎粗壮，直立；基生叶大，叶柄紫红色，叶片心状卵形，边缘皱波状，叶脉紫红色，茎生叶较小；圆锥花序直立，顶生，花3～6枚簇生；花被片6，白绿色或黄绿色；瘦果具3棱，沿棱具翅。花期6～7月。
【分布】华北。
【生境】阔叶林区和山地森林草原区的石质山坡、砾石坡地。
【用途】饲用；药用。

North China rhubarb
Buckwheat family
Perennial herb 30～90 cm tall; stems stout, erect; basal leaves large, petioles purple-red, blades cordate-ovate, margins crisped, veins purple-red, cauline leaves smaller; panicles erect, terminal, flowers in fascicles of 3～6; tepals 6, whitish-green or yellowish-green; achenes trigonous, winged on acies. Flowering June to July.
Distribution: N China
Habitat: Rocky slopes and gravelly hillsides in broadleaf forest areas and montane forest-steppe areas
Use: Forage; medicine

总序大黄
Rheum racemiferum Maxim.
蓼科
Polygonaceae

【特征】多年生草本，高30～70cm；茎直立，中空；基生叶大，叶片革质或近革质，宽卵形至近圆形，基部近心形，主脉常呈紫红色，茎生叶1～3，较小；圆锥花序顶生，花多数，簇生；花被片6，白绿色；瘦果具3棱，沿棱具翅。花期6～7月。

【分布】内蒙古西部、宁夏、甘肃。

【生境】荒漠区石质山坡和干河床。

【用途】饲用。

Racemose rhubarb
Buckwheat family
Perennial herb 30～70 cm tall; stems erect, hollow; basal leaves large, blades leathery or nearly so, broadly ovate to suborbicular, base subcordate, main vein usually purple-red, cauline leaves 1～3, smaller; panicles terminal, flowers numerous, fascicled; tepals 6, whitish-green; achenes trigonous, winged on acies. Flowering June to July.
Distribution: W Inner Mongolia, Ningxia and Gansu
Habitat: Rocky slopes and dry riverbeds in desert areas
Use: Forage

皱叶酸模
Rumex crispus L.
蓼科
Polygonaceae

【特征】多年生草本，高50～100cm；茎直立；叶片披针形或矩圆状披针形，长9～25cm，宽1.5～4cm，基部楔形，边缘皱波状，两面无毛；圆锥花序，花多数，簇生；花被片6，绿色带粉色，内轮花被片具小瘤；瘦果具3棱。花期6～8月。

【分布】东北、华北、西北、华中、四川、云南、广西、福建、台湾；亚洲北部、欧洲、非洲北部、北美洲。

【生境】山地、沟谷、河边、草地。

【用途】饲用；药用。

Curly dock
Buckwheat family
Perennial herb 50～100 cm tall; stems erect; leaf blades lanceolate or oblong-lanceolate, 9～25 cm long, 1.5～4 cm wide, base cuneate, margins crisped, glabrous both sides; panicles with numerous and fascicled flowers; tepals 6, pinkish-green, inner tepals tuberculate; achenes trigonous. Flowering June to August.
Distribution: NE, N, NW and C China, Sichuan, Yunnan, Guangxi, Fujian and Taiwan; N Asia, Europe, N Africa, North America
Habitat: Mountains, ravines, riversides and grasslands
Use: Forage; medicine

毛脉酸模
Rumex gmelinii **Turcz. ex Ledeb.**
蓼科
Polygonaceae

【特征】多年生草本，高30～120cm；茎直立，粗壮；叶片宽大，三角状卵形或三角状心形，全缘或微皱波状，下面脉上被糙毛；圆锥花序直立，有分枝，花多数，簇状轮生；花被片6，绿色带粉色，内轮花被片无小瘤；瘦果具3棱。花期6～8月。
【分布】东北、华北；蒙古、朝鲜、日本、西伯利亚。
【生境】林缘、草甸、河岸。
【用途】饲用；药用。

Hairyvein dock
Buckwheat family
Perennial herb 30～120 cm tall; stems erect, stout; leaf blades broad and large, triangular-ovate or triangular-cordate, entire or repand, hispidulous on veins below; panicles erect, branched, numerous flowers clustered in whorles; tepals 6, pinkish-green, inner tepals without tubercles; achenes trigonous. Flowering June to August.
Distribution: NE and N China; Mongolia, Korea, Japan, Siberia
Habitat: Forest edges, meadows and river banks
Use: Forage; medicine

拍摄人：赵 凡　Photos by Zhao Fan

垫状驼绒藜
Ceratoides compacta (Losinsk.) C. P. Tsien et C. G. Ma
(*Krascheninnikovia compacta* (Losinsk.) Grubov)
藜科
Chenopodiaceae

【特征】垫状半灌木,高5～25cm,全株密被星状毛;分枝密集;叶小,密集,叶柄舟状,叶片椭圆形至矩圆状倒卵形,边缘外卷;雄花花被片4;雌花无花被,果时花管外被短毛。花期6～7月。
【分布】甘肃西南、青海、新疆南部、西藏西北部;塔吉克斯坦。
【生境】海拔3 500～5 500m的亚高山至高山荒漠。
【用途】饲用。

Cushion winterfat
Goosefoot family
Cushion-like subshrub 5～25 cm tall, densely stelipilous throughout; branches dense; leaves small and dense, petioles navicular, blades elliptic to oblong-obovate, margins revolute; staminate flower with 4 tepals; pistillate flowers without tepals, floral tube short-hairy outside in fruit. Flowering June to July.
Distribution: SW Gansu, Qinghai, S Xinjiang and NW Tibet; Tajikistan
Habitat: Subalpine to alpine deserts at 3 500～5 500 m
Use: Forage

尖头叶藜
Chenopodium acuminatum Willd.
藜科
Chenopodiaceae

【特征】一年生草本，高10～80cm；茎直立，分枝平卧或斜升；叶片卵形至菱状卵形，全缘，具半透明的红色或黄褐色环边；花8～10枚，簇生为团伞花序，再组成穗状或圆锥状花序；花被5深裂，背部具隆脊，果时呈五角星状。花期6～8月。

【分布】东北、华北、西北、华东；蒙古、朝鲜、日本、俄罗斯（西伯利亚、远东）、中亚。

【生境】盐碱地、撂荒地、河边、田间。

【用途】饲用；种子可榨油。

Tapertip goosefoot (Pointedleaf goosefoot)
Goosefoot family
Annual herb 10～80 cm tall; stems erect, branches procumbent or ascending; leaf blades ovate to rhombic-ovate, entire, with semi-pellucid and red or yellow-brown margins; flowers 8～10, clustered in a glomerule, further forming a spike or panicle; perianth 5-parted, back keeled, 5-angled in fruit. Flowering June to August.
Distribution: NE, N, NW and E China; Mongolia, Korea, Japan, Russia (Siberia and Far East), Central Asia
Habitat: Saline-alkali sites, abandoned lands, riversides and farmlands
Use: Forage; seeds for extracting oil

刺藜
Chenopodium aristatum L.
(*Dysphania aristata* (L.) Mosyakin et Clemants)
藜科
Chenopodiaceae

- 【特征】一年生草本，高10~40cm，植株淡绿色，秋后变紫红色；茎直立，多分枝；叶片条形至条状披针形，全缘；复二歧式聚伞花序，末端分枝针刺状；花被裂片5，背部稍肥厚，边缘膜质，果时开展。花期8~9月。
- 【分布】东北、华北、西北，山东、河南、四川；蒙古、朝鲜、日本、西伯利亚、中亚、欧洲、北美洲。
- 【生境】沙地、荒地、田间、路边。
- 【用途】饲用；药用。

Thorny goosefoot
Goosefoot family
Annual herb 10~40 cm tall, plants pale green, becoming purplish-red in autumn; stems erect, much branched; leaf blades linear to linear-lanceolate, entire; inflorescence a compound dichasia, apical branchlets acicular; perianth 5-parted, with slightly fleshy back and membranous margins, spreading in fruit. Flowering August to September.
Distribution: NE, N, NW China, Shandong, Henan and Sichuan; Mongolia, Korea, Japan, Siberia, Central Asia, Europe, North America
Habitat: Sandy sites, wastelands, farmlands and roadsides
Use: Forage; medicine

杂配藜
Chenopodium hybridum L.
藜科
Chenopodiaceae

【特征】一年生草本，高40～90cm；茎直立，粗壮，具5锐棱；叶片宽卵形至卵状三角形，掌状浅裂；花数枚团集，排列成疏散的圆锥状花序，顶生或腋生；花被5裂，背部具隆脊，边缘膜质。花期8～9月。
【分布】东北、华北、西北、西南、浙江；蒙古、朝鲜、日本、西伯利亚、中亚、印度、夏威夷群岛、欧洲、北美洲。
【生境】林缘、灌丛、沟谷、河边、村舍附近。
【用途】饲用；药用。

Mapleleaf goosefoot (Sow-bane)
Goosefoot family
Annual herb 40～90 cm tall; stems erect, stout, 5-angled; leaf blades broadly ovate to ovate-triangular, palmately lobed; several flowers glomerulate in a loose panicle, terminal or axillary; perianth 5-parted, with keeled back and membranous margins. Flowering August to September.
Distribution: NE, N, NW and SW China, Zhejiang; Mongolia, Korea, Japan, Siberia, Central Asia, India, Hawaii Islands, Europe, North America
Habitat: Forest margins, thickets, ravines, riversides, residential places
Use: Forage; medicine

拍摄人：赵利清 Photos by Zhao Liqing

碱蓬
Suaeda glauca (Bunge) Bunge
藜科
Chenopodiaceae

【特征】一年生草本，高30～60cm；茎直立，具条纹，上部多分枝；叶半圆柱状或条形，肉质，灰绿色，光滑或被粉粒；团伞花序，花1～5枚着生于叶片基部；花被片5，果时增厚呈五角星状。花期7～8月。
【分布】东北、华北、西北；蒙古、朝鲜、日本、俄罗斯（远东）。
【生境】湿润盐渍化土壤。
【用途】饲用；种子油制做肥皂和油漆；全株富含碳酸钾，为化工原料。

Common seepweed
Goosefoot family
Annual herb 30～60 cm tall; stems erect, striate, much branched above; leaves semi-terete or linear, fleshy, grey-green, smooth or farinose; 1～5 flowers in a glomerule at leaf base; tepals 5, incrassate to 5-angled in fruit. Flowering July to August.
Distribution: NE, N and NW China; Mongolia, Korea, Japan, Russia (Far East)
Habitat: Moist saline soils
Use: Forage; seed oil for making soap and paint; plants rich in potassium carbonate, as raw materials for industrial chemicals

茄叶碱蓬 (阿拉善碱蓬)
Suaeda przewalskii Bunge
藜科
Chenopodiaceae

【特征】一年生草本，高20～40cm，植株绿色，带紫色或紫红色；茎平卧或直立，由基部分枝，枝斜升；叶肉质，略呈倒卵状，先端钝圆，基部渐狭，略弯；花3～10枚簇生于叶腋，呈团伞状；花被片5，肉质，果时背面具狭翅。花期7～9月。

【分布】内蒙古西部、宁夏、甘肃；蒙古。

【生境】荒漠区盐碱湖滨、盐潮洼地、沙丘间低地。

【用途】饲用；全株富含碳酸钾，为化工原料。

Alashan seepweed
Goosefoot family
Annual herb 20～40 cm tall, plants green, purplish to reddish; stems prostrate or erect, branched from base, branches ascending; leaves fleshy, somewhat obovoid, apex rounded, base attenuate, slightly curved; 3～10 flowers glomerulate in leaf axils; tepals 5, fleshy, back narrowly winged in fruit. Flowering July to September.
Distribution: W Inner Mongolia, Ningxia and Gansu; Mongolia
Habitat: Saline-alkali lakeshores, saline swales and bottomlands among dunes in desert areas
Use: Forage; plants rich in potassium carbonate, as raw materials for industrial chemicals

拍摄人：赵利清 Photo by Zhao Liqing

长梗亚欧唐松草
Thalictrum minus var. *stipellatum*
(C. A. Mey. ex Maxim.) Tamura
毛茛科
Ranunculaceae

【特征】多年生草本，高60～120cm；叶为2～4回三出羽状复叶，小叶片先端3浅裂或有疏齿；圆锥花序长达30cm；花梗长10～20mm；萼片4，淡黄绿色，稍带紫色；无花瓣；花药粗于花丝。花期7～8月。
【分布】内蒙古、新疆、亚洲北部、欧洲北部。
【生境】山地草甸、河谷草甸、林下。
【用途】药用。

Longpedicel low meadowrue
Buttercup family
Perennial herb 60～120 cm tall; leaves ternate-pinnate 2～4 times, leaflet blades 3-lobed or spaced-serrate at apex; panicles to 30 cm long; pedicels 10～20 mm long; sepals 4, pale yellowish-green, somewhat purplish; petals without; anthers thicker than the filaments. Flowering July to August.
Distribution: Inner Mongolia and Xinjiang; N Asia, N Europe
Habitat: Montane meadows, valley meadows and forests
Use: Medicine

短梗箭头唐松草
***Thalictrum simplex* var. *brevipes* H. Hara**
毛茛科
Ranunculaceae

【特征】多年生草本，高50~100cm；茎直立；基生叶2至3回羽状复叶，茎生叶向上近直展，2回羽状复叶，上部叶渐变小；圆锥花序顶生；花梗长1~4mm；萼片4，淡黄绿色，略带紫色；无花瓣；花药粗于花丝，黄色。花期7~8月。
【分布】东北、华北、陕西、甘肃、青海、四川、湖北；朝鲜、日本。
【生境】草甸、林缘、灌丛。
【用途】药用。

Shortpedicel arrowhead meadowrue
Buttercup family
Perennial herb 50~100 cm tall; stems erect; basal leaves pinnate twice to thrice, cauline leaves subvertical, bipinnate, gradually reduced upwards; panicles terminal; pedicel 1~4 mm long; sepals 4, pale yellowish-green, somewhat purplish; petals without; anthers thicker than the filaments, yellow. Flowering July to August.
Distribution: NE and N China, Shaanxi, Gansu, Qinghai, Sichuan and Hubei; Korea, Japan
Habitat: Meadows, forest edges and thickets
Use: Medicine

狼毒大戟
Euphorbia fischeriana Steud.
大戟科
Euphorbiaceae

【特征】多年生草本，高20～40cm，内含乳汁；茎直立，单一，无毛；茎基部叶鳞片状，中部叶互生，矩圆形至矩圆状披针形，上部叶3～5轮生，卵状矩圆形；伞幅5～6；腺体5，肾形；蒴果宽卵状，密被短柔毛或变无毛。花期6月。
【分布】东北、华北；蒙古、俄罗斯。
【生境】森林草原及草原。
【用途】有毒。根药用；茎、叶制杀虫剂。

Caper spurge
Spurge family
Perennial herb 20～40 cm tall, with milky juice; stems erect, single, glabrous; basal cauline leaves scale-like, middle leaves alternate, oblong to oblong-lanceolate, upper leaves in whorls of 3～5, ovate-oblong; rays 5～6; glands 5, reniform; capsules broadly ovoid, densely pubescent or glabrate. Flowering June.
Distribution: NE and N China; Mongolia, Russia
Habitat: Forest-steppe and steppe
Use: Poisonous. Roots for medicine; stems and leaves for making insecticide

刘氏大戟
Euphorbia lioui C. Y. Wu et J. S. Ma
大戟科
Euphorbiaceae

【特征】多年生草本，高约15cm，内含乳汁；茎直立，中部以上多分枝；叶互生，条形至倒卵状披针形；总苞叶4～5，卵状披针形；伞幅4～5；腺体4，褐色，边缘齿裂。花期5月。
【分布】内蒙古西部特有。
【生境】荒漠地带的山前平原。
【 注 】有毒。

Liu spurge
Spurge family
Perennial herb about 15 cm tall, with milky juice; stems erect, much branched in half above; leaves alternate, linear to obovate-lanceolate; involucral leaves 4～5, ovate-lanceolate; rays 4～5; glands 4, brown, dentate-lobed. Flowering May.
Distribution: Endemic to W Inner Mongolia
Habitat: Plains in front of mountains in desert zone
Note: Poisonous

拍摄人：赵利清 Photos by Zhao Liqing

拍摄人：卢欣石　Photos by Lu Xinshi

准噶尔大戟
Euphorbia soongarica Boiss.
大戟科
Euphorbiaceae

【特征】多年生草本，高50～100cm，内含乳汁，全株光滑无毛；茎丛生，直立；叶互生，倒披针形至狭矩圆形；伞幅3～5，苞叶2，卵形至矩圆形，黄色；腺体5，半圆形，橙黄色至淡褐色；蒴果近球状，光滑或疏生疣点。花期6～8月。
【分布】新疆北部、甘肃西部；中亚、蒙古、西伯利亚。
【生境】荒漠地带的河谷、盐化草甸、山坡、田边、路旁。
【用途】有毒。根药用。

Dzungar spurge
Spurge family
Perennial herb 50～100 cm tall, with milky juice, glabrous throughout; stems tufted, erect; leaves alternate, oblanceolate to narrowly oblong; rays 3～5, bracteal leaves 2, ovate to oblong, yellow; glands 5, semi-rounded, orange-yellow to pale brown; capsules subglobose, smooth or sparsely verrucose. Flowering June to August.
Distribution: N Xinjiang and W Gansu; Central Asia, Mongolia, Siberia
Habitat: Valleys, saline meadows, slopes, farmland sides and roadsides in desert zone
Use: Poisonous. Roots for medicine

一叶萩
Flueggea suffruticosa (Pall.) Baill.
大戟科
Euphorbiaceae

【特征】灌木，高1～3m；多分枝，枝光滑无毛；单叶互生，全缘或具齿，两面光滑；花雌雄异株，雄花3～18枚簇生于叶腋，雌花单一或数枚簇生于叶腋；萼片5，黄绿色；无花瓣；蒴果3棱状扁球形，具网纹。花期6～7月。
【分布】除甘肃、青海、新疆和西藏外，全国各地；蒙古、朝鲜、日本、俄罗斯。
【生境】石质山坡、山地灌丛、林缘、路边。
【用途】药用；纺织原料；提取栲胶。

Fountain hardhack
Spurge family
Shrubs 1～3 m tall; branches many, glabrous; simple leaves alternate, entire or serrate, smooth; flowers dioecious, staminate flowers 3～18 clustered in axils, pistillate flowers solitary or several clustered in axils; sepals 5, yellowish-green; petals without; capsules triquetrous-oblate, reticulate-veined. Flowering June to July.
Distribution: Throughout China except Gansu, Qinghai, Xinjiang and Tibet; Mongolia, Korea, Japan, Russia
Habitat: Rocky slopes, montane scrublands, forest edges and roadsides
Use: Medicine; raw materials for textile; extracting tannin

拍摄人：赵利清 Photos by Zhao Liqing

拍摄人：赵利清　Photos by Zhao Liqing

地构叶
Speranskia tuberculata (Bunge) Baill.
大戟科
Euphorbiaceae

【特征】多年生草本，高20～50cm；茎直立；单叶互生，披针形或卵状披针形，边缘具疏齿或深裂；花雌雄同株；总状花序顶生；萼片5，淡绿色，疏被长柔毛；花瓣5，鳞片状；腺体小；蒴果3瓣裂，被柔毛和瘤状突起。花期6月。
【分布】东北、华北、西北、华东。
【生境】石质山坡、草坡、灌丛。
【用途】根有毒。药用。

Rough speranskia
Spurge family
Perennial herb 20～50 cm tall; stems erect; simple leaves alternate, lanceolate or ovate-lanceolate, margins sparsely toothed or parted; flowers monoecious; racemes terminal; sepals 5, pale green, sparsely pilose; petals 5, scale-like; glands tiny; capsules 3-valved, pilose and verrucose. Flowering June.
Distribution: NE, N, NW and E China
Habitat: Rocky and grassy slopes, thickets
Use: Root poisonous. Medicine

柳叶鼠李
Rhamnus erythroxylum Pall.
鼠李科
Rhamnaceae

【特征】灌木，高达2m；多分枝，枝先端针刺状；叶条状披针形，边缘具疏细齿，齿端具黑色腺点；花黄绿色，10～20枚簇生于短枝上；萼片5；花瓣5；核果球形，熟时黑色，具2～3核；每核具2～3粒种子。花期5月。

【分布】华北，陕西、甘肃；蒙古、西伯利亚。

【生境】山坡、沙丘间低地。

【用途】饲用；叶药用。

Willowleaf buckthorn
Buckthorn family
Shrub to 2 m tall; branches many, ending in a thorn; leaves linear-lanceolate, margins spaced-serrulate with black glands; flowers yellow-green, 10～20 clustered on short branchlets; sepals 5; petals 5; drupes globose, black at maturity, with 2 or 3 stones; 2- or 3-seeded per stone. Flowering May.

Distribution: N China, Shaanxi and Gansu; Mongolia, Siberia

Habitat: Mountain slopes and lowlands among dunes

Use: Forage; leaves for medicine

小叶鼠李
Rhamnus parvifolia **Bunge**
鼠李科
Rhamnaceae

- 【特征】灌木，高达2m；多分枝，枝先端针刺状；叶质厚，菱状卵形、倒卵形或椭圆形，边缘具疏细齿，齿端具黑色腺点，下面脉腋腺窝具柔毛；花黄绿色，1～3枚簇生于叶腋；萼片4；花瓣4；核果球形，熟时黑色，具2核；每核具1粒种子。花期5月。
- 【分布】华北、辽宁、山东、甘肃；蒙古、朝鲜、西伯利亚。
- 【生境】干旱石质山坡、沙丘间低地。
- 【用途】饲用；果实药用；水土保持；庭院绿化。

Littleleaf buckthorn
Buckthorn family
Shrub to 2 m tall; branches many, ending in a thorn; leaves thick, rhombic-ovate, obovate or elliptic, margins spaced-serrulate with black glands, vein axils with puberulent areole beneath; flowers yellow-green, 1～3 clustered in an axil; sepals 4; petals 4; drupes globose, black at maturity, with 2 stones; 1-seeded per stone. Flowering May.
Distribution: N China, Liaoning, Shandong and Gansu; Mongolia, Korea, Siberia
Habitat: Dry rocky slopes and lowlands among dunes
Use: Forage; fruits for medicine; soil conservation; courtyard planting

酸枣
Ziziphus jujuba var. *spinosa* (Bunge) Hu ex H. F. Chow
鼠李科
Rhamnaceae

【特征】灌木或小乔木,高达4m;小枝略呈"之"字形,具刺;单叶互生,边缘有齿,齿端具腺点;花小,2~3枚簇生于叶腋,花瓣5,黄绿色;核果卵状至矩圆状,长7~15mm,暗红色,熟后变黑色;核顶端钝。花期5~6月。
【分布】东北、华北、西北、华中、华东;俄罗斯。
【生境】山地阳坡或山谷。
【用途】绿篱;水土保持;蜜源;茎皮可提制栲胶;果实可提取维生素或酿酒;核壳可制活性炭;种子可榨油;全株药用。

Sour jujube
Buckthorn family
Shrub or small tree to 4 m tall; twigs somewhat zigzagged, armed; simple leaves alternate, margins toothed with glands; small flowers 2 or 3 clustered in an axil; petals 5, yellow-green; drupes ovoid to oblong, 7~15 mm long, dark red, becoming black at maturity; stones blunt at apex. Flowering May to June.
Distribution: NE, N, NW, C and E China; Russia
Habitat: Mountain sunny slopes and valleys
Use: Green fences; soil conservation; honey source; barks for extracting tannin; fruits for extracting vitamin and wine-brewing; stone shells for making active carbon; seeds for extracting oil; whole plant for medicine

车前
Plantago asiatica L.
车前科
Plantaginaceae

【特征】二年生或多年生草本；须根；叶基生，宽卵形至宽椭圆形，基部近圆形或宽楔形，边缘波状、具疏齿或全缘；花葶直立或弓曲上升，长20～50cm；穗状花序圆柱形；花具短梗；花冠裂片4，长约1mm，干膜质，反折。花期6～8月。

【分布】几遍全国；亚洲、欧洲。

【生境】草甸、沟谷、田野、田间、路边。

【用途】饲用；药用。

Asian plantain
Plantain family

Biennial or perennial herb; fibrous roots; leaves basal, broadly ovate to broadly elliptic, base suborbicular or broadly cuneate, margins undulate, spaced-serrate or entire; scapes erect or bent-ascending, 20～50 cm long; spikes cylindric; flowers short-pedicellate; corolla with 4 lobes about 1 mm long, scarious, reflexed. Flowering June to August.

Distribution: Almost throughout China; Asia, Europe

Habitat: Meadows, gullies, fields, farmlands and roadsides

Use: Forage; medicine

平车前
Plantago depressa Willd.
车前科
Plantaginaceae

【特征】一年生或二年生草本；直根圆柱状；叶基生，椭圆形至披针形，基部狭楔形，边缘具疏齿或全缘；花葶直立或弓曲上升，长5～20cm；穗状花序圆柱形；花冠裂片4，长0.5～1mm，干膜质，反折。花期6～7月。
【分布】几遍全国；蒙古、朝鲜、日本、俄罗斯（西伯利亚、远东）、中亚、巴基斯坦、印度。
【生境】草甸、田野、路边、居民点附近。
【用途】饲用；药用；嫩叶可食；种子可提炼工业用油。

Depressed plantain
Plantain family
Annual or biennial herb; taproot terete; leaves basal, elliptic to lanceolate, base narrowly cuneate, margins spaced-serrate or entire; scapes erect or bent-ascending, 5～20 cm long; spikes cylindric; corolla with 4 lobes to 0.5～1 mm long, scarious, reflexed. Flowering June to July.
Distribution: Almost throughout China; Mongolia, Korea, Japan, Russia (Siberia and Far East), Central Asia, Pakistan, India
Habitat: Meadows, fields, roadsides and residential places
Use: Forage; medicine; young leaves edible; seeds for extracting industrial oil

条叶车前
Plantago lessingii Fisch. et C. A. Mey.
车前科
Plantaginaceae

【特征】一年生草本，高3～20cm，全株密被长柔毛；叶基生，条形，平卧，全缘；花葶斜升或直立；穗状花序卵形至矩圆形；花冠裂片4，干膜质，反折。花期6～8月。
【分布】西北，内蒙古，山西；蒙古，中亚。
【生境】荒漠至草原带山地、丘陵、沟谷。
【用途】饲用。

Linearleaf plantain
Plantain family
Annual herb 3～20 cm tall, densely villous throughout; leaves basal, linear, procumbent, entire; scapes ascending or erect; spikes ovoid to oblong; corolla with 4 lobes, scarious, reflexed. Flowering June to August.
Distribution: NW China, Inner Mongolia and Shanxi; Mongolia, Central Asia
Habitat: Mountains, hills and ravines in desert to steppe zones
Use: Forage

大车前
Plantago major L.
车前科
Plantaginaceae

【特征】二年生或多年生草本；须根；叶基生，宽卵形至宽椭圆形，基部近圆形或宽楔形，边缘波状、具疏齿或近全缘；花葶直立或弓曲上升，长6～40cm；穗状花序圆柱形；花无梗；花冠裂片4，长1～1.5mm，干膜质，反折。花期6～8月。

【分布】几遍全国；欧亚大陆温带及寒温带。

【生境】草甸、沼泽、山坡、沟谷、渠边、田边、路旁。

【用途】饲用；药用。

Common plantain
Plantain family
Biennial or perennial herb; fibrous roots; leaves basal, broadly ovate to broadly elliptic, base suborbicular or broadly cuneate, margins undulate, spaced-serrate or subentire; scapes erect or bent-ascending, 6～40 cm long; spikes cylindric; flowers sessile; corolla with 4 lobes to 1～1.5 mm long, scarious, reflexed. Flowering June to August.
Distribution: Almost throughout China; temperate to cool-temperate zones in Eurasia
Habitat: Meadows, swamps, slopes, gullies, ditch sides, farmland sides and roadsides
Use: Forage; medicine

拍摄人：赵 凡　Photos by Zhao Fan

盐生车前
Plantago maritima var. *salsa* (Pall.) Pilger
车前科
Plantaginaceae

【特征】多年生草本；叶基生，条形或狭条形，直立或平卧，全缘；花葶直立或弓曲上升，长5～40cm；穗状花序圆柱形；花冠裂片4，长约1.5mm，干膜质，反折。花期6～8月。
【分布】西北、河北、内蒙古、陕西；蒙古、西伯利亚、中亚、阿富汗、伊朗、高加索。
【生境】盐化草甸、盐湖边缘、戈壁。
【用途】饲用。

Goose tongue (Sea plantain, Seaside plantain)
Plantain family
Perennial herb; leaves basal, linear or narrowly linear, erect or procumbent, entire; scapes erect or bent-ascending, 5～40 cm long; spikes cylindric; corolla with 4 lobes about 1.5 mm long, scarious, reflexed. Flowering June to August.
Distribution: NW China, Hebei, Inner Mongolia and Shaanxi; Mongolia, Siberia, Central Asia, Afghanistan, Iran, Caucasia
Habitat: Saline meadows, salt lake edges and gobi
Use: Forage

拍摄人：卢欣石　Photos by Lu Xinshi

拍摄人：吴新宏　Photos by Wu Xinhong

海韭菜
Triglochin maritima L.
水麦冬科
Juncaginaceae

【特征】多年生草本；叶基生，条形，稍肉质，光滑；花葶直立，高20～50cm；总状花序，花排列较紧密；花梗长约1mm；花被片6，绿色；雌蕊6；蒴椭圆状或卵状，成熟后呈6瓣开裂。花期6月。
【分布】东北、华北、西北、西南；广布于北半球温带和寒带。
【生境】河湖边盐渍化草甸、海边盐滩。
【用途】药用。植株含氢氰酸，对牲畜有毒。

Seaside arrowgrass
Arrowgrass family
Perennial herb; leaves basal, linear, somewhat fleshy, smooth; scapes erect, 20～50 cm tall; racemes with flowers relatively crowded; pedicels about 1 mm long; tepals 6, green; pistils 6; capsules ellipsoid or ovoid, 6-valved at maturity. Flowering June.
Distribution: NE, N, NW and SW China; widespread in temperate to frigid zones in the Northern Hemisphere
Habitat: Saline meadows along riversides and lakeshores, salty swales along seashores
Use: Medicine. Plants containing hydrocyanic acid, poisonous to livestock

水麦冬
Triglochin palustris L.
水麦冬科
Juncaginaceae

【特征】多年生草本；叶基生，条形；花葶直立，高20～60cm；总状花序，花排列较疏松；花梗长约2mm；花被片6，绿色带紫色；雌蕊3；蒴果棒状条形，成熟后呈3瓣开裂。花期6月。
【分布】东北、华北、西北、西南；广布于北半球温带和寒带。
【生境】咸湿地、浅水边。
【用途】药用。植株含氢氰酸，对牲畜有毒。

Marsh arrowgrass
Arrowgrass family
Perennial herb; leaves basal, linear; scapes erect, 20～60 cm tall; racemes with flowers relatively distributed; pedicels about 2 mm long; tepals 6, purplish-green; pistils 3; capsules clavate-linear, 3-valved at maturity. Flowering June.
Distribution: NE, N, NW and SW China; widespread in temperate to frigid zones in the Northern Hemisphere
Habitat: Brackish marshes and shallow watersides
Use: Medicine. Plants containing hydrocyanic acid, poisonous to livestock

拍摄人：赵利清 Photos by Zhao Liqing

巨序剪股颖
Agrostis gigantea Roth
禾本科
Poaceae (Gramineae)

【特征】多年生草本，高60～120cm；具根茎；秆丛生，直立或基部膝曲；叶片扁平，叶舌长5～6mm；圆锥花序开展；小穗含1小花；两颖近等长；外稃短于颖，无芒；内稃长为外稃的3/4。花期6～7月。

【分布】东北、华北、西北、西南、长江流域；欧亚大陆温带广布。

【生境】草甸、林缘、沟边、溪旁。

【用途】饲用。

Redtop (Black bent grass)
Grass family
Perennial 60～120 cm tall; rhizomatous; culms tufted, erect or geniculate at base; leaf blades flat, ligules 5～6 mm long; panicles open; spikelets with 1 floret; glumes subequal; lemma shorter than the glumes, awnless; palea 3/4 length of the lemma. Flowering June to July.
Distribution: NE, N, NW and SW China, and regions along Changjiang River; widespread in temperate zone in Eurasia
Habitat: Meadows, forest margins, ditch sides and streamsides
Use: Forage

光稃香草
Anthoxanthum glabrum (Trin.) Veldkamp
(光稃茅香 *Hierochloe glabra* Trin.)
禾本科
Poaceae (Gramineae)

【特征】多年生草本，高10～30cm，植株有香味；具根茎；秆直立；叶片扁平；圆锥花序开展；小穗有光泽，含3小花，顶生小花两性，两侧生小花雄性；颖膜质；两性小花外稃披针形。花期7～8月。
【分布】东北、华北、华东，青海、新疆、云南；蒙古、俄罗斯（西伯利亚、远东）、哈萨克斯坦。
【生境】湿润草地。
【用途】饲用。

Smoothglume sweetgrass
Grass family
Perennial 10～30 cm tall, fragrant; rhizomatous; culms erect; leaf blades flat; panicles open; spikelets lustrous, with 3 florets, the terminal one bisexual, the lateral 2 staminate; glumes membranous; lemma of bisexual floret lanceolate. Flowering July to August.
Distribution: NE, N and E China, Qinghai, Xinjiang and Yunnan; Mongolia, Russia (Siberia and Far East), Kazakhstan
Habitat: Moist grasslands
Use: Forage

拍摄人：赵利清 Photos by Zhao Liqing

右图拍摄人：拾 涛 Right photos by Shi Tao

三芒草
Aristida adscensionis L.
禾本科
Poaceae (Gramineae)

【特征】一年生草本，高5～60 cm；秆丛生，直立、斜升或平卧；叶片内卷；圆锥花序，分枝单生；小穗含1小花；颖长5～10mm；外稃长6.5～12mm，顶端具3芒，芒粗糙无毛。花期6～8月。

【分布】东北、华北、西北、河南、山东；蒙古、印度、阿富汗、中亚、高加索、小亚细亚、地中海沿岸。

【生境】荒漠至荒漠草原地带的干燥坡地、丘陵、浅沟、干河床、沙地。

【用途】饲用；固沙。

Sixweeks threeawn
Grass family
Annual 5~60 cm tall; culms caespitose, erect, ascending or procumbent; leaf blades involute; panicles with single branche per node; spikelets with 1 floret; glumes 5～10 mm long; lemmas 6.5～12 mm long, 3-awned at apex, awns scabrous and glabrous. Flowering June to August.
Distribution: NE, N and NW China, Henan and Shandong; Mongolia, India, Afghanistan, Central Asia, Caucasia, Asia minor, areas along the Mediterranean Sea
Habitat: Dry slopes, hills, shallow ditches, dry riverbeds and sands in desert to desert-steppe zones
Use: Forage; fixing dunes

大颖三芒草
Aristida grandiglumis **Roshev.**
(大颖针禾 *Stipagrostis grandiglumis* (Roshev.) Tzvelev)
禾本科
Poaceae (Gramineae)

【特征】多年生草本，高30~65cm；秆密丛生，直立；叶片内卷；圆锥花序开展，分枝单生；小穗含1小花；颖不等长，第一颖长25~30mm，第二颖长20~23mm；外稃长8~9mm，顶端微2裂，无毛，具3芒，芒全体密被羽状柔毛。花期6~8月。
【分布】甘肃（敦煌）、新疆南部；蒙古。
【生境】沙漠。
【用途】饲用；固沙。

Largeglume threeawn
Grass family
Perennial 30~65 cm tall; culms strongly caespitose, erect; leaf blades involute; panicles open, with single branche per node; spikelets with 1 floret; glumes unequal, the lower 25~30 mm long, the upper 20~23 mm long; lemmas 8~9 mm long, apex slightly 2-lobed, glabrous, 3-awned, awns densely plumose throughout. Flowering June to August.
Distribution: Gansu (Dunhuang) and S Xinjiang; Mongolia
Habitat: Sandy desert
Use: Forage; fixing dunes

羽毛三芒草
Aristida pennata Trin.
(羽毛针禾 *Stipagrostis pennata* (Trin.) De Winter)
禾本科
Poaceae (Gramineae)

【特征】多年生草本，高20～60cm；秆丛生，直立；叶片内卷；圆锥花序疏松，分枝单生或孪生；小穗含1小花；颖近等长，长10～17mm；外稃长5～7mm，顶端截形，有短纤毛，具3芒，芒全体密被羽状柔毛。花期7～8月。

【分布】新疆；阿富汗、中亚、伊朗、高加索。

【生境】荒漠地带的沙地或沙丘。

【用途】饲用；固沙。

Feather threeawn
Grass family
Perennial 20~60 cm tall; culms caespitose, erect; leaf blades involute; panicles loose, with single or paired branches per node; spikelets with 1 floret; glumes subequal, 10～17 mm long; lemmas 5～7 mm long, apex truncate, ciliolate, 3-awned, awns densely plumose throughout. Flowering July to August.
Distribution: Xinjiang; Afghanistan, Central Asia, Iran, Caucasia
Habitat: Sands or dunes in desert zone
Use: Forage; fixing dunes

下图拍摄人：拾 涛 Lower photo by Shi Tao

荩草
Arthraxon hispidus (Thunb.) Mak.
禾本科
Poaceae (Gramineae)

【特征】一年生草本，高20～40cm；秆细弱，基部倾斜；叶鞘具疣毛；叶片卵状披针形，基部心形抱茎；2至多数总状花序指状排列，或簇生于茎顶；有柄小穗退化；无柄小穗含1小花，外稃透明膜质，芒自近基部伸出。花期7～8月。
【分布】全国各地；亚、欧、非洲温暖地区。
【生境】水边湿地、低湿草甸、沟谷、田边。
【用途】饲用；药用；黄色染料。

Small carpgrass (Joint-head grass, Joint-head arthraxon)
Grass family
Annual 20~40 cm tall; culms slender, decumbent at base; leaf sheaths tuberculate-hispid; leaf blades ovate-lanceolate, base cordate-clasping; 2 to several racemes digitately arranged or fascicled at culm apex; pediceled spikelets rudimentary; sessile spikelets with 1 floret, lemmas hyaline-membranous, awns arising near base. Flowering July to August.
Distribution: Throughout China; temperate and warm regions in Asia, Europe and Africa
Habitat: Watersides, moist meadows in lowlands, gullies and farmland sides
Use: Forage; medicine; yellow dyestuff

野燕麦
Avena fatua L.
禾本科
Poaceae (Gramineae)

【特征】一年生草本，高60～150cm；秆直立；叶片扁平；圆锥花序开展；小穗含2～3小花；颖边缘膜质；外稃革质，通常下半部被长硬毛，芒长2～4cm，自背部伸出，膝曲，芒柱扭转；颖果黄褐色，成熟后不易与稃片分离。花期4～8月。

【分布】全国各地；亚、欧、非洲温寒带地区。

【生境】山坡、林缘、田边、荒地。

【用途】饲用；药用。

Wild oats
Grass family
Annual 60～150 cm tall; culms erect; leaf blades flat; panicles open; spikelet with 2~3 florets; glumes with membranous margins; lemmas leathery, usually hispid in half below, awns 2～4 cm long, arising from back, geniculate, column twisted; caryopsis yellow-brown, not free from lemma and palea at maturity. Flowering April to August.
Distribution: Throughout China; temperate and cool regions in Asia, Europe and Africa
Habitat: Slopes, forest edges, field sides and wastelands
Use: Forage; medicine

右图拍摄人：拾 涛 Right photo by Shi Tao

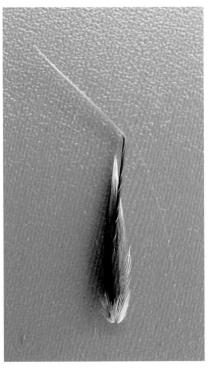

白羊草
***Bothriochloa ischaemum* (L.) Keng**
禾本科
Poaceae (Gramineae)

- 【特征】多年生草本，高30～60cm；具根茎；秆丛生，直立或基部膝曲；叶舌具纤毛，叶片扁平或边缘稍内卷；3至多枚总状花序呈伞房状兼指状排列于主轴上；小穗孪生；无柄小穗第二外稃具芒，芒长10～15mm；有柄小穗无芒。花期7～8月。
- 【分布】全国各地；全球暖温带地区。
- 【生境】草地、灌丛、阔叶林。
- 【用途】饲用；药用；水土保持。

Yellow bluestem (King ranch bluestem, Plains bluestem)
Grass family
Perennial 30～60 cm tall; rhizomatous; culms tufted, erect or geniculate at base; ligules ciliate, leaf blades flat or slightly involute; 3 to several racemes corymbosely or digitately arranged at rachis; spikelets twinning; the second lemma of sessile spikelets with an awn to 10～15 mm long; pediceled spikelets awnless. Flowering July to August.
Distribution: Throughout China; warm-temperate zone in the world
Habitat: Grasslands, scrublands and broadleaf woodlands
Use: Forage; medicine; soil conservation

缘毛雀麦
Bromus ciliatus L.
禾本科
Poaceae (Gramineae)

【特征】多年生草本，高60～120cm；具短根茎；秆直立，节常被倒柔毛；叶片扁平；圆锥花序于花期开展，分枝弯曲；小穗含3～10小花；外稃边缘膜质，中部以下边缘被柔毛，芒长2～6mm。花期7～8月。
【分布】东北，内蒙古；蒙古、俄罗斯（远东）、北美。
【生境】林缘草地、路旁、沟边。
【用途】饲用。

Fringed brome
Grass family
Perennial 60～120 cm tall; short-rhizomatous; culms erect, joints usually retrorse-pubescent; leaf blades flat; panicles open in flower, with curved branches; spikelets with 3~10 florets; lemmas marginally membranous and pilose in lower half, awns 2～6 mm long. Flowering July to August.
Distribution: NE China, Inner Mongolia; Mongolia, Russia (Far East), North America
Habitat: Meadows along forest margins, roadsides and gully sides
Use: Forage

沙地雀麦
Bromus ircutensis Kom.
禾本科
Poaceae (Gramineae)

- 【特征】多年生草本，高50～90cm；具根茎；秆直立；叶鞘通常被绒毛，叶片扁平；圆锥花序直立，收缩，分枝被毛；小穗含5～12小花；颖膜质；外稃先端钝圆，无芒，基部边缘被柔毛。花期7～8月。
- 【分布】内蒙古（浑善达克沙地）；蒙古、西伯利亚。
- 【生境】草原区固定或半固定沙丘。
- 【用途】饲用；固沙。

Sandbinding brome (Irkutsk brome)
Grass family
Perennial 50～90 cm tall; rhizomatous; culms erect; leaf sheaths usually tomentose, blades flat; panicles erect, contracted, with pubescent branches; spikelets with 5~12 florets; glumes membranous; lemmas blunt-rounded at apex, awnless, base marginally pilose. Flowering July to August.
Distribution: Inner Mongolia (Hunshandake Sands); Mongolia, Siberia
Habitat: Fixed and semi-fixed dunes in steppe areas
Use: Forage; fixing dunes

假苇拂子茅
Calamagrostis pseudophragmites **(Haller f.) Koeler**
禾本科
Poaceae (Gramineae)

【特征】多年生草本，高30～100cm；具根茎；秆直立；叶片扁平或内卷；圆锥花序疏松开展，长10～20（35）cm，分枝簇生，斜升；小穗含1小花；颖成熟后张开；外稃透明膜质，芒长约3mm，自近顶端处伸出。花期7～8月。
【分布】东北、华北、西北、四川、云南、贵州、湖北；欧亚大陆温带广布。
【生境】山坡草地、河岸、沟谷、低地、沙地。
【用途】饲料；水土保持。

False common reedgrass
Grass family
Perennial 30～100 cm tall; rhizomatous; culms erect; leaf blades flat or involute; panicles loosely open, 10～20 (35) cm long, with clustered and ascending branches; spikelets with 1 floret; glumes open at maturity; lemmas hyaline-membranous, awns about 3 mm long, arising near the top. Flowering July to August.
Distribution: NE, N and NW China, Sichuan, Yunnan, Guizhou and Hubei; widespread in temperate zone in Eurasia
Habitat: Sloping grasslands, river banks, gullies, lowlands and sands
Use: Forage; soil conservation

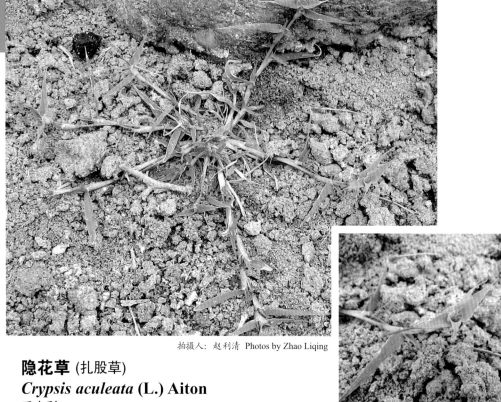

拍摄人：赵利清 Photos by Zhao Liqing

隐花草 (扎股草)
Crypsis aculeata (L.) Aiton
禾本科
Poaceae (Gramineae)

【特征】一年生草本；秆铺散、平卧或斜升，长5～40cm；叶鞘疏松，膨大；叶片条状披针形，扁平或对折，边缘多少内卷；圆锥花序短缩呈头状，下面托以2枚苞片状叶鞘；小穗含1小花；颖狭窄，膜质；外稃宽于颖，膜质，无芒。花期7～8月。

【分布】华东、华北、陕西、宁夏、甘肃、新疆；蒙古、俄罗斯、中亚、西亚、南欧、北非。

【生境】河岸、沟旁、盐碱地。

【用途】饲用。

Common pricklegrass
Grass family
Annual; culms diffuse, procumbent or ascending, 5～40 cm long; leaf sheaths loose and inflated; leaf blades linear-lanceolate, flat or folded, margins somewhat involute; panicles shortened into a head, subtended by 2 bracteiform sheaths; spikelets with 1 floret; glumes narrow, membranous; lemmas wider than glumes, membranous, awnless. Flowering July to August.

Distribution: E and N China, Shaanxi, Ningxia, Gansu and Xinjiang; Mongolia, Russia, Central Asia, W Asia, S Europe, N Africa

Habitat: River banks, ditch sides, saline-alkali sites

Use: Forage

狗牙根
Cynodon dactylon (L.) Pers.
禾本科
Poaceae (Gramineae)

【特征】多年生草本，具根茎和匍匐茎；直立茎高10～30cm，匍匐茎长6～100cm；叶片扁平，叶舌具纤毛；穗状花序3～6枚呈指状着生于茎顶端；小穗含1小花；颖背部成脊，边缘膜质；外稃舟形。花期6～8月。
【分布】黄河以南各省区，新疆；世界温暖地区。
【生境】河岸、渠边、田边、路旁、平原绿洲。
【用途】饲用；坪用；水土保持；药用。世界上广泛用于草坪建植。

Bermudagrass (Dog-tooth grass)
Grass family
Perennial; rhizomatous and stoloniferous; standing culms 10~30 cm tall, stolons 6～100 cm long; leaf blades flat, ligules ciliate; 3～6 spikes digitately arranged at culm apex; spikelets with 1 floret; glumes keeled, margins membranous; lemmas navicular. Flowering June to August.
Distribution: Regions to south of Yellow River, and Xinjiang; warm regions in the world
Habitat: River banks, ditch sides, field sides, roadsides and plain oasis
Use: Forage; lawn; soil conservation; medicine. Widely used as a lawn grass in the world

拍摄人：吴新宏 Photos by Wu Xinhong

鸭茅
***Dactylis glomerata* L.**
禾本科
Poaceae (Gramineae)

【特征】多年生草本，高40～200cm；秆直立或基部膝曲；叶片扁平，质软；圆锥花序开展，分枝单生或基部孪生；小穗含2～5小花；颖脊上粗糙或具纤毛；外稃脊上粗糙或具纤毛，先端具约1mm的短芒。花期5～8月。
【分布】西南，新疆；欧亚大陆温带。
【生境】林缘、草甸。
【用途】饲用。作为优良牧草在世界上温带和亚热带地区广泛栽培。

Orchardgrass (Cock's-foot grass)
Grass family
Perennial 40～200 cm tall; culms erect or geniculate at base; leaf blades flat and soft; panicles open, branches single or basally paired; spikelets with 2～5 florets; glumes scabrous or ciliolate on rib; lemmas scabrous or ciliolate on rib, apex with a short awn about 1 mm. Flowering May to August.
Distribution: SW China, and Xinjiang; temperate zone in Eurasia
Habitat: Forest edges and meadows
Use: Forage. As an excellent forage widely cultivated in temperate and subtropical regions in the world

毛马唐
Digitaria ciliaris var. *chrysoblephara*
(Fig. et De Not.) R. R. Stewart
禾本科
Poaceae (Gramineae)

【特征】一年生草本，高30～100cm；秆基部倾卧；叶鞘常具柔毛，叶片扁平；总状花序4至10枚呈指状排列；小穗排列于穗轴一侧，孪生；第一外稃具长柔毛和刚毛。花期6～8月。
【分布】东北、华北、河南、陕西、甘肃、四川、安徽、江苏；世界亚热带和温带地区。
【生境】田野、路边。
【用途】饲用。

Hairy crabgrass
Grass family
Annual 30～100 cm tall; culms decumbent; leaf sheaths usually pilose, leaf blades flat; 4～10 racemes digitately arranged; spikelets arranged on one side of rachis, twinning; the first lemma pilose and setose. Flowering June to August.
Distribution: NE and N China, Henan, Shaanxi, Gansu, Sichuan, Anhui and Jiangsu; subtropic and temperate regions in the world
Habitat: Fields and roadsides
Use: Forage

拍摄人：赵利清 Photos by Zhao Liqing

阿拉善披碱草
Elymus alashanicus (Keng ex Keng et S. L. Chen) S. L. Chen
(阿拉善鹅观草 *Roegneria alashanica* Keng ex Keng et S. L. Chen)
禾本科
Poaceae (Gramineae)

【特征】多年生草本，高40～70cm；有时具根茎；秆疏丛生，直立或基部斜升；叶片内卷，坚韧，直立；穗状花序直立，细瘦，穗轴每节具1小穗；小穗含4～6小花，小穗轴平滑；颖平滑无毛，边缘膜质；外稃平滑无毛，无芒。花期6～7月。
【分布】内蒙古中部和西部、宁夏、甘肃、新疆。
【生境】石质山坡、岩崖、石缝。
【用途】饲用。

Alashan wildrye
Grass family
Perennial 40～70 cm tall; sometimes rhizomatous; culms loosely tufted, erect or basally ascending; leaf blades involute, stiff and straight; spikes erect, slender, with 1 spikelet per node; spikelets with 4~6 florets, rachilla smooth; glumes glabrous, with membranous margins; lemmas glabrous, awnless. Flowering June to July.
Distribution: C and W Inner Mongolia, Ningxia, Gansu and Xinjiang
Habitat: Rocky slopes, rock cliffs and rock crevices
Use: Forage

圆柱披碱草
Elymus dahuricus var. *cylindricus* Franch.
(*Elymus cylindricus* (Franch.) Honda)
禾本科
Poaceae (Gramineae)

【特征】多年生草本，高40～80cm；秆疏丛生，直立或基部膝曲；叶片扁平，宽约5mm，干后内卷；穗状花序直立，穗轴每节具2小穗；小穗含2～3小花；颖先端具芒，芒长2～4mm；外稃被短小糙毛，第一外稃长7～8mm，芒直立或稍外展，长6～13mm。花期7～8月。
【分布】河北、内蒙古、陕西、宁夏、青海、新疆、四川、云南、河南。
【生境】草甸。
【用途】饲用。

Cylindric wildrye
Grass family
Perennial 40～80 cm tall; culms loosely tufted, erect or geniculate at base; leaf blades flat, about 5 mm wide, involute when dry; spikes erect, with 2 spikelets per node; spikelets with 2～3 florets; glumes with an awn to 2～4 mm long; lemmas hirsutulous, the first lemma 7～8 mm long, awns straight or slightly extrorse, 6～13 mm long. Flowering July to August.
Distribution: Hebei, Inner Mongolia, Shaanxi, Ningxia, Qinghai, Xinjiang, Sichuan, Yunnan and Henan
Habitat: Meadows
Use: Forage

肥披碱草
Elymus excelsus Turcz. ex Griseb.
禾本科
Poaceae (Gramineae)

【特征】多年生草本，高60～140cm；秆粗壮，丛生，直立或基部膝曲；叶片扁平，宽10～16mm；穗状花序直立，穗轴每节具2～3（4）小穗；小穗含4～5小花；颖先端具芒，芒长约7mm；外稃先端、脉上和边缘被微小短毛，第一外稃长8～12mm，芒粗糙，反曲，长15～20（40）mm。花期6～8月。

【分布】东北、华北、西北、四川、河南；蒙古、朝鲜、日本、俄罗斯（西伯利亚、远东）。

【生境】草甸。

【用途】饲用。

Tall wildrye
Grass family
Perennial 60～140 cm tall; culms stout, caespitose, erect or geniculate at base; leaf blades flat, 10～16 mm wide; spikes erect, with 2～3 (4) spikelets per node; spikelets with 4～5 florets; glumes with an awn about 7 mm long; lemmas with puberulent apex, veins and margins, the first lemma 8～12 mm long, awns scabrous and reflexed, 15～20 (40) mm long. Flowering June to August.

Distribution: NE, N and NW China, Sichuan and Henan; Mongolia, Korea, Japan, Russia (Siberia and Far East)
Habitat: Meadows
Use: Forage

麦薲草
Elymus tangutorum (Nevski) Hand.-Mazz.
禾本科
Poaceae (Gramineae)

【特征】多年生草本，高60～120cm；秆疏丛生，直立，基部膝曲；叶片扁平；穗状花序直立，较紧密，穗轴每节具2小穗；小穗有时稍偏于一侧，绿色，略带紫色，含3～4小花；颖先端尖或具1～3mm短芒；外稃无毛或上半部被微小硬毛，芒直立，长3～11mm。花期7～8月。
【分布】华北、西北、西南。
【生境】林间空地、灌丛、草甸、田边。
【用途】饲用。

Tangut wildrye
Grass family
Perennial 60～120 cm tall; culms loosely tufted, erect, geniculate at base; leaf blades flat; spikes erect, relatively dense, with 2 spikelets per node; spikelets sometimes slightly secund, purplish-green, with 3~4 florets; glumes pointed or with an awn to 1～3 mm; lemmas glabrous or hirsutulous in upper half, awns straight, 3～11 mm long. Flowering July to August.
Distribution: N, NW and SW China
Habitat: Forest openings, thickets, meadows and farmland sides
Use: Forage

野黍
Eriochloa villosa (Thunb.) Kunth
禾本科
Poaceae (Gramineae)

【特征】一年生草本，高30~100cm；秆直立或稍倾斜；叶片扁平；总状花序4~8枚排列于主穗轴一侧，密生柔毛；小穗单生，成2行排列于穗轴一侧；第一颖退化，第二颖膜质，被毛；第一外稃膜质，被毛，第二外稃革质。花期7~8月。
【分布】东北、华北、华东、华中、华南、西南；日本、俄罗斯、印度、伊朗。
【生境】田边、路旁、水边等潮湿处。
【用途】饲用；种子食用。

Hairy cupgrass (Woolly cupgrass)
Grass family
Annual 30~100 cm tall; culms erect or slightly ascending; leaf blades flat; 4~8 racemes arranged on one side of rachis, densely pubescent; spikelet single, secund in 2 rows; the lower glume vestigial, the upper glume membranous and pubescent; the first lemma membranous and pubescent, the second one leathery. Flowering July to August.
Distribution: NE, N, E, C, S and SW China; Japan, Russia, India, Iran
Habitat: Moist places along field sides, roadsides and watersides
Use: Forage; seeds edible

达乌里羊茅
Festuca dahurica (St.-Yves) V. I. Krecz. et Bobrov
禾本科
Poaceae (Gramineae)

【特征】多年生草本，高30～60cm；秆密丛生，直立；叶片内卷，秆生叶长2～3cm，宽0.6～1mm；圆锥花序较紧缩，长6～8cm，花序轴及分枝被短柔毛；小穗绿色或淡紫色，含4～6小花；颖平滑；外稃被细短毛或粗糙，先端尖，无芒。花期6月。
【分布】东北，内蒙古、河北、甘肃；蒙古、西伯利亚。
【生境】草原带沙地、石质山坡。
【用途】饲用。

Dahurian fescue
Grass family
Perennial 30～60 cm tall; culms strongly caespitose, erect; leaf blades involute, culm leaf blades 2～3 cm long, 0.6～1 mm wide; panicles relatively contracted, 6～8 cm long, rachis and branches pubescent; spikelets green or pale purple, with 4~6 florets; glumes smooth; lemmas puberulent or scabrous, apex pointed, awnless. Flowering June.
Distribution: NE China, Inner Mongolia, Hebei and Gansu; Mongolia, Siberia
Habitat: Sands and rocky slopes in steppe zone
Use: Forage

紫羊茅
Festuca rubra L.
禾本科
Poaceae (Gramineae)

【特征】多年生草本，高20～60cm；具根茎；秆疏丛生，直立；叶片对折或内卷，或秆生叶扁平；圆锥花序狭窄，花期开展；小穗绿色或紫色，含5～7小花；颖平滑或微粗糙；外稃平滑、粗糙或被细短毛，芒长1.5～3mm。花期6月。
【分布】东北、华北、西北、西南、华中；北半球温带。
【生境】海拔600～4 500m的河谷至山地或高山带的草甸、林缘及灌丛。
【用途】饲用；坪用。

Red fescue
Grass family
Perennial 20～60 cm tall; rhizomatous; culms loosely tufted, erect; leaf blades folded or involute, or culm leaf blades flat; panicles narrow, open in flower; spikelets green or purple, with 5~7 florets; glumes smooth or scaberulose; lemmas smooth, scabrous or puberulent, awns 1.5～3 mm long. Flowering June.
Distribution: NE, N, NW, SW and C China; temperate zone in the Northern Hemisphere
Habitat: Meadows, forest margins, thickets in valleys to montane or alpine belts at 600～4 500 m
Use: Forage; lawn

异燕麦
Helictotrichon schellianum (Hack.) Kitag.
禾本科
Poaceae (Gramineae)

【特征】多年生草本，高40～80cm；秆疏丛生，直立；叶片扁平或稍内卷，叶舌长3～6mm；圆锥花序较紧缩；小穗有光泽，含3～6小花；颖上端膜质；外稃上部透明膜质，芒长12～15mm，自背部伸出。花期7～8月。
【分布】东北、华北、西北、西南；蒙古、西伯利亚、中亚。
【生境】山地草原、草甸。
【用途】饲用。

Schelle spikeoat
Grass family
Perennial 40～80 cm tall; culms loosely tufted, erect; leaf blades flat or slightly involute, ligules 3～6 mm long; panicles relatively contracted; spikelets lustrous, with 3~6 florets; glumes apically membranous; lemmas apically hyaline-membranous, awns 12～15 mm long, arising from back. Flowering July to August.
Distribution: NE, N, NW and SW China; Mongolia, Siberia, Central Asia
Habitat: Montane steppe, meadows
Use: Forage

左图拍摄人：拾 涛 Left photo by Shi Tao

臭草
Melica scabrosa Trin.
禾本科
Poaceae (Gramineae)

【特征】多年生草本，高20～70cm；秆丛生，直立或基部膝曲；叶片扁平；圆锥花序狭窄，分枝直立或斜升，小穗密集；小穗含2～6枚能育小花；颖几等长，狭披针形，长4～7mm；外稃背部粗糙。花期5～6月。
【分布】东北、华北、西北、华东、华中、西南；蒙古、朝鲜。
【生境】山坡草地、田野、渠边、路旁。
【用途】饲用，但家畜采食过多可中毒。

Rough melic
Grass family
Perennial 20～70 cm tall; culms caespitose, erect or geniculate at base; leaf blades flat; panicles narrow, branches erect or ascending, with dense spikelets; spikelets with 2～6 fertile florets; glumes subequal, narrowly lanceolate, 4～7 mm long; lemmas scabrous. Flowering May to June.
Distribution: NE, N, NW, E, C and SW China; Mongolia, Korea
Habitat: Sloping grasslands, fields, ditch sides and roadsides
Use: Forage, but poisonous to livestock when taking too much

大臭草
Melica turczaninowiana **Ohwi**
禾本科
Poaceae (Gramineae)

【特征】多年生草本，高70～130cm；秆丛生，直立或基部膝曲；叶片扁平；圆锥花序开展，分枝斜升或开展；小穗含2～3枚能育小花；颖几等长，卵状矩圆形，长9～11mm；外稃背部中部以下被糙毛。花期6～8月。
【分布】东北、华北；蒙古、俄罗斯（西伯利亚和远东）。
【生境】山地林缘、林地、灌丛和草甸。
【注　】对家畜有毒。

Big melic
Grass family
Perennial 70～130 cm tall; culms caespitose, erect or geniculate at base; leaf blades flat; panicles open, branches ascending or spreading; spikelets with 2～3 fertile florets; glumes subequal, ovate-oblong, 9～11mm long; lemmas hispidulous in lower half. Flowering June to August.
Distribution: NE and N China; Mongolia, Russia (Siberia and Far East)
Habitat: Montane forest margins, woodlands, scrublands and meadows
Note: Poisonous to livestock

拍摄人：赵利清　Photos by Zhao Liqing

抱草
Melica virgata Turcz. ex Trin.
禾本科
Poaceae (Gramineae)

【特征】多年生草本，高30~70cm；秆丛生，直立或基部膝曲；叶片常内卷；圆锥花序细长，分枝直立或斜升；小穗含2~3枚能育小花；颖不等长，第一颖卵形，长2~3mm；外稃背部粗糙，有时疏被糙毛。花期7~8月。
【分布】华北；蒙古、西伯利亚。
【生境】山坡草地和灌丛。
【用途】饲用，但家畜采食过多可中毒。

Slender melic
Grass family
Perennial 30~70 cm tall; culms caespitose, erect or geniculate at base; leaf blades usually involute; panicles slender, branches erect or ascending; spikelets with 2~3 fertile florets; glumes unequal, the lower one ovate, 2~3 mm long; lemmas scabrous, or sparsely hispidulous. Flowering July to August.
Distribution: N China; Mongolia, Siberia
Habitat: Sloping grasslands and scrublands
Use: Forage, but poisonous to livestock when taking too much

虉草
Phalaris arundinacea L.
禾本科
Poaceae (Gramineae)

【特征】多年生草本,高60~150cm;具根茎;秆直立或基部膝曲;叶片扁平;圆锥花序紧密;小穗含两性小花1枚及不孕外稃2枚;颖脊粗糙,有时具狭翼;孕花外稃宽披针形,有光泽。花期6~7月。
【分布】东北、华北、西北、华东、华中;北半球温带地区广布。
【生境】潮湿草地和水湿处。
【用途】饲用;编织;造纸。

Reed canarygrass
Grass family
Perennial 60~150 cm tall; rhizomatous; culms erect or geniculate at base; leaf blades flat; panicles contracted and dense; spikelets with 1 perfect floret and 2 sterile lemmas; glume ribs scabrous, sometimes narrowly winged; fertile lemma broadly lanceolate, lustrous. Flowering June to July.
Distribution: NE, N, NW, E and C China; widespread in temperate zone in the Northern Hemisphere
Habitat: Wet meadows and watersides
Use: Forage; weaving; papermaking

拍摄人:张洪江 Photos by Zhang Hongjiang

芦苇
Phragmites australis (Cav.) Trin. ex Steud.
禾本科
Poaceae (Gramineae)

【特征】多年生草本，高10~250cm；具根茎；秆直立、斜升或倾卧，坚硬；叶片扁平，光滑或边缘粗糙；圆锥花序稠密，开展；小穗长12~16mm，含3~5枚小花；颖先端锐尖；第一小花常为雄性，两性小花外稃先端长渐尖，基盘被长柔毛。花期7~9月。
【分布】全国各地；世界广布。
【生境】河湖边水湿处、潮湿草地、盐碱地、沙地、坡地。
【用途】饲用；造纸；编织；药用。

Common reed
Grass family

Perennial 10~250 cm tall; rhizomatous; culms erect, ascending or procumbent, rigid; leaf blades flat, smooth or marginally scabrous; panicles dense and open; spikelets 12~16 mm long, with 3~5 florets; glumes acute at apex; the lowest floret usually male, bisexual lemmas long-acuminate at apex, callus pilose. Flowering July to September.
Distribution: Throughout China; cosmopolitan
Habitat: Watersides along rivers and lakes, wet meadows, saline-alkali sites, sands and slopes
Use: Forage; papermaking; weaving; medicine

细叶早熟禾
Poa pratensis ssp. *angustifolia* (L.) Lejeun
(*Poa angustifolia* L.)
禾本科
Poaceae (Gramineae)

【特征】多年生草本，高30~60cm；具根茎；秆直立，丛生，光滑；分蘖叶常内卷，秆生叶对折或扁平；圆锥花序，分枝直立或上举；小穗含2~5小花；外稃脊下部2/3及边脉下部1/2被长柔毛，基盘密生长绵毛。花期6~7月。
【分布】东北、华北、西北、西南；北半球温带广布。
【生境】海拔500~4 400m的林缘、草甸、沟谷、坡地。
【用途】饲用；坪用。

Narrowleaf bluegrass
Grass family
Perennial 30~60 cm tall; rhizomatous; culms erect, tufted, smooth; basal leaves usually involute, culm leaves folded or flat; panicles with erect or ascending branches; spikelets with 2~5 florets; lemmas villous on lower 2/3 of keel and lower 1/2 of marginal veins, callus densely long-lanate. Flowering June to July.
Distribution: NE, N, NW and SW China; widespread in temperate zone in the Northern Hemisphere
Habitat: Forest edges, meadows, gullies and slopes at 500~4 400 m
Use: Forage; lawn

左图拍摄人：拾 涛 Left photo by Shi Tao

假泽早熟禾
Poa pseudopalustris Keng
禾本科
Poaceae (Gramineae)

【特征】多年生草本，高40～80cm；具短根茎；秆直立或基部稍膝曲；叶片扁平；圆锥花序开展，先端稍下垂；小穗含2～4小花；外稃椭圆形，先端尖而狭膜质，脊及边脉被柔毛，基盘密被绵毛。花期6～7月。
【分布】东北，内蒙古东部、青海。
【生境】林缘、草甸。
【用途】饲用。

False marsh bluegrass
Grass family
Perennial 40～80 cm tall; short-rhizomatous; culms erect or slightly geniculate at base; leaf blades flat; panicles open, slightly nodding at apex; spikelets with 2～4 florets; lemmas elliptic, apex pointed and narrowly membranous, pubescent on keel and marginal veins, callus densely lanate. Flowering June to July.
Distribution: NE China, E Inner Mongolia and Qinghai
Habitat: Forest edges and meadows
Use: Forage

光稃早熟禾
Poa psilolepis Keng ex L. Liou
禾本科
Poaceae (Gramineae)

【特征】多年生草本，高30～60cm；秆密丛生，直立或基部稍膝曲；叶片内卷，质硬；圆锥花序狭窄稍疏松；小穗含2～4小花；颖披针形；外稃披针形，无毛，先端膜质，基盘无绵毛。花期8～10月。

【分布】青海、甘南、川西。

【生境】海拔3 100～4 200m的草甸、灌丛、疏林。

【用途】饲用。

Smoothlemma bluegrass
Grass family
Perennial 30～60 cm tall; culms strongly caespitose, erect or slightly geniculate at base; leaf blades involute, rigid; panicles narrow and relatively loose; spikelets with 2～4 florets; glumes lanceolate; lemmas lanceolate, glabrous, apex membranous, callus not lanate. Flowering August to October.
Distribution: Qinghai, S Gansu and W Sichuan
Habitat: Meadows, thickets and open woodlands at 3 100～4 200 m
Use: Forage

左图拍摄人：拾 涛 Left photo by Shi Tao

硬质早熟禾
Poa sphondylodes Trin.
禾本科
Poaceae (Gramineae)

【特征】多年生草本，高20～60cm；秆密丛生，直立，近花序处稍粗糙；叶片扁平；圆锥花序紧缩；小穗含3～6小花；颖披针形；外稃披针形，脊下部2/3及边脉下部1/2被柔毛，基盘被适量绵毛。花期6月。
【分布】东北、华北、西北、山东、江苏；蒙古、日本、西伯利亚。
【生境】草原、灌丛、草甸。
【用途】饲用；药用。

Hard bluegrass
Grass family
Perennial 20～60 cm tall; culms strongly caespitose, erect, scaberulose below inflorescence; leaf blades flat; panicles contracted; spikelets with 3～6 florets; glumes lanceolate; lemmas lanceolate, pubescent on lower 2/3 of keel and lower 1/2 of marginal veins, callus moderately lanate. Flowering June.
Distribution: NE, N and NW China, Shandong and Jiangsu; Mongolia, Japan, Siberia
Habitat: Steppe, thickets and meadows
Use: Forage; medicine

中亚细柄茅
Ptilagrostis pelliotii (Danguy) Grub.
禾本科
Poaceae (Gramineae)

【特征】多年生草本，高20～50cm；秆密丛生，直立；叶片质硬，粗糙，内卷，叶舌截平，被纤毛；圆锥花序疏松开展；颖膜质，光滑无毛；外稃遍生柔毛，芒长20～25mm，遍生长柔毛。花期6～7月。
【分布】西北，内蒙古西部；蒙古、中亚。
【生境】荒漠地带的戈壁滩、砾石质坡地、岩缝。
【用途】饲用。

Pelliot false needlegrass
Grass family
Perennial 20～50 cm tall; culms strongly caespitose, erect; leaf blades rigid, scabrous, involute, ligules truncate, ciliate; panicles loosely open; glumes membranous, glabrous; lemmas pubescent throughout, awns 20～25 mm long, villous throughout. Flowering June to July.
Distribution: NW China, W Inner Mongolia; Mongolia, Central Asia
Habitat: Gobi, gravelly slopes and rock crevices in desert zone
Use: Forage

碱茅
Puccinellia distans (Jacq.) Parl.
禾本科 Poaceae (Gramineae)

【特征】多年生草本，高15～50cm；秆丛生，直立或基部膝曲；叶片扁平或内卷；圆锥花序开展，下部分枝通常于成熟后水平开展或下伸；小穗含3～7小花；颖先端钝，第一颖长约1mm；外稃先端钝或截平，基部被短毛；花药长0.5～0.8mm。花期5～6月。
【分布】华北；欧亚大陆温带、北美洲。
【生境】碱化低湿地和草甸。
【用途】饲用。

Weeping alkaligrass
Grass family
Perennial 15～50 cm tall; culms tufted, erect or geniculate at base; leaf blades flat or involute; panicles open, lower branches usually horizontally spreading or declined at maturity; spikelets with 3～7 florets; glumes obtuse at apex, lower glume about 1 mm long; lemmas obtuse or truncate at apex, pubescent at base; anthers 0.5～0.8 mm long. Flowering May to June.
Distribution: N China; temperate zone in Eurasia, North America
Habitat: Alkali moist lowlands and meadows
Use: Forage

金色狗尾草
Setaria glauca (L.) Beauv.
禾本科
Poaceae (Gramineae)

【特征】一年生草本，高20～90cm；秆直立、倾斜或基部膝曲；叶片扁平，叶舌具纤毛；圆锥花序紧密，圆柱状；刚毛金黄色，长6～8mm。花期6～8月。
【分布】全国各地；欧亚大陆温带至热带地区。
【生境】田间、路旁、荒地。
【用途】饲用；药用；种子可食用。

Yellow foxtail
Grass family
Annual 20～90 cm tall; culms erect, ascending or geniculate at base; leaf blades flat, ligules ciliate; panicles contracted and dense, cylindric; bristles golden, 6～8 mm long. Flowering June to August.
Distribution: Throughout China; temperate to tropic zones in Eurasia
Habitat: Farmlands, roadsides and wastelands
Use: Forage; medicine; seeds edible

狗尾草
Setaria viridis (L.) P. Beauv.
禾本科
Poaceae (Gramineae)

【特征】一年生草本，高20～60cm；秆直立或基部稍膝曲；叶片扁平，叶舌具纤毛；圆锥花序紧密，圆柱状；刚毛绿色、黄色或带紫色，长于小穗2～4倍。花期7～9月。
【分布】全国各地；全球温带至热带地区。
【生境】草地、荒地、田间、路旁。
【用途】饲用；药用；种子可食用

Green bristlegrass
Grass family
Annual 20～60 cm tall; culms erect or slightly geniculate at base; leaf blades flat, ligules ciliate; panicles contracted and dense, cylindric; bristles green, yellow or purplish, 2 to 4 times longer than spikelets. Flowering July to September.
Distribution: Throughout China; temperate to tropic zones in the world
Habitat: Grasslands, wastelands, farmlands and roadsides
Use: Forage; medicine; seeds edible

异针茅
Stipa aliena Keng
禾本科
Poaceae (Gramineae)

【特征】多年生草本，高20～40cm；秆丛生，直立；叶片内卷；圆锥花序较紧缩；外稃背部遍生短毛；芒二回膝曲，第一芒柱长4～5mm，具长柔毛，第二芒柱与第一芒柱几等长，被微毛，芒针长10～16mm，无毛。花期7～8月。
【分布】内蒙古（阿拉善）、甘肃、青海、四川、西藏。
【生境】海拔2 900～4 600m的山坡、冲积扇、河谷阶地。
【用途】饲用。

Strange needlegrass
Grass family
Perennial 20～40 cm tall; culms caespitose, erect; leaf blades involute; panicles relatively contracted; lemmas short-hairy throughout; awns 2-geniculate, column 4～5 mm to the first bend with villi, similar length to the second bend with puberulence, bristle 10～16 mm long, glabrous. Flowering July to August.
Distribution: Inner Mongolia (Alashan), Gansu, Qinghai, Sichuan and Tibet
Habitat: Mountain slopes, alluvial fans and valley terraces at 2 900～4 600 m
Use: Forage

拍摄人：赵利清 Photos by Zhao Liqing

长芒草 (本氏针茅)
Stipa bungeana Trin.
禾本科
Poaceae (Gramineae)

【特征】多年生草本,高30~60cm;秆密丛生,直立或斜升;叶片内卷;圆锥花序每节具2~4分枝;芒二回膝曲,光滑或粗糙,第一芒柱长1~1.5cm,第二芒柱长0.5~1cm,芒针长3~5cm。花期6~7月。
【分布】东北至西北及西南;蒙古、中亚。
【生境】山地、丘陵、沟谷。中国北方暖温型草原建群种。
【用途】饲用。

Tufted needlegrass
Grass family
Perennial 30~60 cm tall; culms strongly caespitose, erect or ascending; leaf blades involute; panicles with 2~4 branches per node; awns 2-geniculate, glabrous or scabrous throughout, column 1~1.5 cm to the first bend, 0.5~1 cm to the second bend, bristle 3~5 cm long. Flowering June to July.
Distribution: NE to NW and SW China; Mongolia, Central Asia
Habitat: Mountains, hills and ravines. A dominant species in warm-temperate steppe in northern China
Use: Forage

龙须菜
Asparagus schoberioides Kunth
百合科
Liliaceae

【特征】多年生草本，高40～100cm；根状茎粗壮；茎直立，分枝斜升；叶状枝2～6簇生，条形，基部近三棱形，具中脉；花2～4枚腋生；花梗极短；花冠钟形，黄绿色；浆果球形，熟时红色。花期6～7月。
【分布】东北、华北、山东、河南、陕西、甘肃；蒙古、朝鲜、日本、俄罗斯。
【生境】林地、林缘和草甸。
【用途】饲用；根状茎和根药用。

Shooting asparagus
Lily family
Perennial herb 40～100 cm tall; rhizomes stout; stems erect, branches ascending; cladophylls in fascicles of 2～6, linear, nearly 3-angled at base, midnerved; flowers 2～4, axillary; pedicels very short; corolla campanulate, yellowish-green; berries globose, red at maturity. Flowering June to July.
Distribution: NE and N China, Shandong, Henan, Shaanxi and Gansu; Mongolia, Korea, Japan, Russia
Habitat: Forests, forest edges and meadows
Use: Forage; rhizomes and roots for medicine

拍摄人：赵利清 Photos by Zhao Liqing

拍摄人：赵利清 Photos by Zhao Liqing

曲枝天门冬
Asparagus trichophyllus Bunge
百合科
Liliaceae

【特征】多年生草本，高20～70cm；茎近直立，中部以上强烈廻折状，分枝先下弯而后上升；叶状枝5～10簇生，刚毛状；花1～2枚腋生；花梗长12～16mm；花冠钟形，黄绿色；浆果球形，熟时紫红色。花期6～7月。
【分布】辽宁、山东、河北、内蒙古中部；蒙古、西伯利亚。
【生境】山坡草地、灌丛、荒地、路边。
【用途】饲用。

Threadleaf asparagus
Lily family
Perennial herb 20～70 cm tall; stems suberect, strongly flexuose in upper half, branches basally curved downward, distally ascending; cladophylls in fascicles of 5～10, setiform; flowers 1 or 2, axillary; pedicels 12～16 mm long; corolla campanulate, yellowish-green; berries globose, purple-red at maturity. Flowering June to July.
Distribution: Liaoning, Shandong, Hebei and C Inner Mongolia; Mongolia, Siberia
Habitat: Sloping grasslands, scrublands, wastelands and roadsides
Use: Forage

圆叶木蓼
Atraphaxis tortuosa Losinsk.
蓼科
Polygonaceae

【特征】灌木，高50～60cm，多分枝；叶革质，近圆形至宽卵形，边缘具皱波状齿，两面密被腺点；总状花序顶生；花被片5，粉色或白色；瘦果尖卵状，具3棱。花期5～6月。
【分布】内蒙古；蒙古。
【生境】半荒漠地区的石质低山、丘陵。
【用途】饲用；水土保持。

Roundleaf knotwood
Buckwheat family
Shrub 50～60 cm tall, much branched; leaves leathery, suborbicular to broadly ovate, margins crisped-serrate, densely glandular both sides; racemes terminal; tepals 5, pink or white; achenes acute-ovoid, trigonous. Flowering May to June.
Distribution: Inner Mongolia; Mongolia
Habitat: Rocky low mountains and hills in semi-desert areas
Use: Forage; soil conservation

拍摄人：赵利清 Photos by Zhao Liqing

上图拍摄人：赵利清 Upper photo by Zhao Liqing

木藤首乌
Fallopia aubertii (L. Henry) Holub
(木藤蓼 *Polygonum aubertii* L. Henry)
蓼科
Polygonaceae

【特征】半灌木；茎缠绕；叶片矩圆状卵形或卵形，近革质，基部近心形；花序圆锥状，少分枝，稀疏；花被白色或淡绿色，5深裂，外轮3片较大，背部具翅；瘦果具3棱。花期6~7月。
【分布】华中、西南、内蒙古（贺兰山）、山西、陕西、宁夏、甘肃、青海。
【生境】海拔900~3 200m的山地林缘、山谷灌丛。
【用途】饲用；药用；庭院绿化。

Mile-a-minute vine (Chinese fleece vine, Fleece flower, Silver lace vine)
Buckwheat family
Subshrub; stems twining; leaf blades oblong-ovate or ovate, subleathery, base subcordate; inflorescence paniculate, few branched, loose; perianth white or greenish, 5-parted, the outer 3 larger and winged; achenes trigonous. Flowering June to July.
Distribution: C and SW China, Inner Mongolia (Helan Mountains), Shanxi, Shaanxi, Ningxia, Gansu and Qinghai
Habitat: Montane forest margins and walley thickets at 900~3 200 m
Use: Forage; medicine; courtyard planting

两栖蓼
Polygonum amphibium L.
蓼科
Polygonaceae
【特征】多年生草本,水陆两生;水生者茎横走,无毛,叶浮于水面,具长柄,叶片无毛;陆生者茎直立或斜升,被长硬毛,叶具短柄或近无柄,叶片被硬毛;花序穗状,顶生;花被粉红色或白色,5深裂;瘦果双凸镜状。花期6~9月。
【分布】全国各地;几遍北半球温带。
【生境】河边、湖滨、沟渠、低湿地、农田。
【用途】饲用;药用。

Water smartweed
Buckwheat family
Perennial herb, aquatic and terrestrial; aquatic stems transverse, glabrous, leaves floating, long-petiolate, blades glabrous; terrestrial stems erect or ascending, hirsute, leaves short-petiolate or subsessile, blades strigose; inflorescence spiciform, terminal; perianth pink or white, 5-parted; achenes lenticular. Flowering June to September.
Distribution: Throughout China; almost throughout the northern temperate zone
Habitat: Riversides, lakeshores, ditches, moist lowlands and farmlands
Use: Forage; medicine

左图拍摄人:吴新宏 Left photo by Wu Xinhong

珠芽蓼
Polygonum viviparum L.
蓼科
Polygonaceae

【特征】多年生草本,高10~35cm;根茎肥厚;茎直立,不分枝;基生叶与茎下部叶具长柄,上部叶无柄;总状花序穗状,顶生,下半部生珠芽;苞片卵形,膜质;花被白色或粉红色,5深裂;瘦果卵状,具3棱。花期6~7月。

【分布】东北、华北、西北、西南;蒙古、朝鲜、日本、哈萨克斯坦、印度、高加索、欧洲、北美洲。

【生境】海拔1 200~5 100m的山地及亚高山和高山带的林缘、灌丛、草甸。

【用途】饲用;根茎药用并提取栲胶;珠芽及根茎食用。

Alpine bistort
Buckwheat family
Perennial herb 10~35 cm tall; rhizomes thick; stems erect, unbranched; basal and lower cauline leaves long-petiolate, upper leaves sessile; racemes spiciform, terminal, bearing bulblets in half below; bracts ovate, membranous; perianth white or pink, 5-parted; achenes ovoid, trigonous. Flowering June to July.
Distribution: NE, N, NW and SW China; Mongolia, Korea, Japan, Kazakhstan, India, Caucasia, Europe, North America
Habitat: Forest edges, thickets and meadows in montane to subalpine and alpine belts at 1 200~5 100 m
Use: Forage; rhizomes for medicine, and extracting tannin; bulblets and rhizomes edible

灯心草蚤缀
Arenaria juncea M. Bieb.
石竹科
Caryophyllaceae

【特征】多年生草本,高20~50cm;茎直立,丛生,上部被腺毛;叶丝状,坚硬;二歧聚伞花序顶生;花梗与萼片被腺毛;花瓣5,白色,先端圆。花期6~8月。
【分布】东北、华北、西北;蒙古、朝鲜、日本、俄罗斯(西伯利亚、远东)。
【生境】石质山坡和丘顶。
【用途】饲用;药用。

Rush sandwort
Pink family
Perennial herb 20~50 cm tall; stems erect, caespitose, glandular-hairy above; leaves filiform, stiff; dichasium terminal; pedicels and sepals glandular-hairy; petals 5, white, rounded at apex. Flowering June to August.
Distribution: NE, N and NW China; Mongolia, Korea, Japan, Russia (Siberia and Far East)
Habitat: Rocky slopes and hilltops
Use: Forage; medicine

卷耳
Cerastium arvense L.
石竹科
Caryophyllaceae

【特征】多年生草本,高10～30cm;茎疏丛生,直立,密被柔毛,上部混生腺毛;叶条状披针形至矩圆状披针形,被柔毛或混生腺毛;二歧聚伞花序顶生;花梗与萼片密被腺毛;花瓣5,白色,先端2浅裂;花柱5。花期5～7月。
【分布】东北、华北、西北;蒙古、日本、俄罗斯、北美洲。
【生境】山地林缘、草甸、溪边。
【用途】饲用。

Field chickweed (Meadow chickweed)
Pink family
Perennial herb 10～30 cm tall; stems loosely caespitose, erect, densely pubescent, mixed with glandular-hairs above; leaves linear-lanceolate to oblong-lanceolate, pubescent or mixed with glandular-hairs; dichasium terminal; pedicels and sepals densely glandular-hairy; petals 5, white, 2-lobed at apex; styles 5. Flowering May to July.
Distribution: NE, N and NW China; Mongolia, Japan, Russia, North America
Habitat: Montane forest edges, meadows and streamsides
Use: Forage

拍摄人：赵 凡（左图）& 吴新宏（右图） Photos by Zhao Fan & Wu Xinhong

旱麦瓶草 (山蚂蚱草)
***Silene jenisseensis* Willd.**
石竹科
Caryophyllaceae

【特征】多年生草本，高20~50cm；茎丛生，直立或斜升，不分枝，无毛；基生叶簇生，披针状条形，茎生叶少数；狭聚伞状圆锥花序顶生或腋生；萼筒无毛，脉间白色膜质；花瓣5，白色至淡绿色，2裂。花期6~8月。

【分布】东北、华北；蒙古、朝鲜、俄罗斯（西伯利亚、远东）。

【生境】草原、沙地、草甸、林缘。

【用途】饲用；药用。

Aridland catchfly (Mountain grasshopper plant)
Pink family
Perennial herb 20~50 cm tall; stems caespitose, erect or ascending, unbranched, glabrous; basal leaves clustered, lanceolate-linear, cauline leaves few; narrow thyrse terminal or axillary; calyx tube glabrous, internerve white-membranous; petals 5, white to greenish, bifid. Flowering June to August.
Distribution: NE and N China; Mongolia, Korea, Russia (Siberia and Far East)
Habitat: Steppe, sands, meadows and forest margins
Use: Forage; medicine

拍摄人：赵 凡 Photos by Zhao Fan

毛萼麦瓶草 (蔓茎蝇子草)
Silene repens Patrin ex Pers.
石竹科
Caryophyllaceae

【特征】多年生草本，高15~50cm；茎直立或斜升，有分枝，被短柔毛；叶条形至条状倒披针形；狭聚伞状圆锥花序顶生；萼筒常带紫色，密被短柔毛；花瓣5，白色、黄白色至绿白色，2裂。花期6~8月。

【分布】东北、华北、西北，四川、西藏；蒙古、朝鲜、日本、俄罗斯（西伯利亚、远东）、中亚、欧洲。

【生境】草甸、林地、林缘、溪边、沙丘。

【用途】饲用；药用。

Pink campion
Pink family
Perennial herb 15~50 cm tall; stems erect or ascending, branched, pubescent; leaves linear to linear-oblanceolate; narrow thyrse terminal; calyx tube usually purplish, densely pubescent; petals 5, white, yellowish to greenish, bifid. Flowering June to August.
Distribution: NE, N and NW China, Sichuan and Tibet; Mongolia, Korea, Japan, Russia (Siberia and Far East), Central Asia, Europe
Habitat: Meadows, woodlands, forest margins, streamsides and dunes
Use: Forage; medicine

狗筋麦瓶草 (白玉草)
Silene venosa (Gilib.) Asch.
石竹科
Caryophyllaceae

【特征】多年生草本，高40～100cm，全株呈灰绿色；茎丛生，直立，上部分枝；叶披针形至卵状披针形；聚伞花序大型；萼筒膜质，膨大成囊泡状，网脉略带紫色，无毛；花瓣5，白色，2裂。花期6～8月。

【分布】黑龙江、内蒙古东部；蒙古、俄罗斯（西伯利亚、远东）、中亚、尼泊尔、印度、伊朗、土耳其、欧洲、北非、南美洲和北美洲。

【生境】河谷草甸、路边。

【用途】饲用；药用；根可做肥皂代用品。

Maiden's tears
Pink family
Perennial herb 40～100 cm tall, plants grey-green throughout; stems caespitose, erect, branched above; leaves lanceolate to ovate-lanceolate; cymes large; calyx tube membranous, inflated to bladder-like with purplish net-veins, glabrous; white petals 5, bifid. Flowering June to August.
Distribution: Heilongjiang and E Inner Mongolia; Mongolia, Russia (Siberia and Far East), Central Asia, Nepal, India, Iran, Turkey, Europe, N Africa, South America, North America
Habitat: River valley meadows and roadsides
Use: Forage; medicine; roots for cleansing

拍摄人：赵 凡 Photos by Zhao Fan

沙地繁缕
Stellaria gypsophiloides Fenzl
石竹科
Caryophyllaceae
【特征】多年生草本，高30～60cm，全株密被腺毛或柔毛；茎多数，从基部多次二歧分枝，枝缠结交错，形成球状草丛；叶条形至椭圆形；聚伞花序分枝多，呈大型圆锥状；花瓣5，白色，2深裂，裂片条形。花期7～8月。
【分布】内蒙古中部、陕西、宁夏；蒙古。
【生境】沙丘、沙地。
【用途】饲用；固沙；根药用。

Baby breath chickweed
Pink family
Perennial herb 30～60 cm tall, densely piloglandulose or pubescent throughout; stems many, dichotomous-branched several times from base, branches intricate into a globose thicket; leaves linear to elliptic; inflorescence large paniculate, formed from much branched cymes; petals 5, white, parted into 2 linear segments. Flowering July to August.
Distribution: C Inner Mongolia, Shaanxi and Ningxia; Mongolia
Habitat: Dunes and sands
Use: Forage; fixing dunes; roots for medicine

短尾铁线莲
Clematis brevicaudata DC.
毛茛科
Ranunculaceae

【特征】藤本；叶为1至2回三出或羽状复叶，小叶先端长渐尖，边缘具缺刻状牙齿，有时3裂；圆锥状聚伞花序腋生或顶生；萼片4，开展，白色或带黄色，外面沿边缘密被短柔毛；无花瓣；瘦果多数,宿存花柱羽毛状。花期8～9月。
【分布】东北、华北、西北、华东、西南；蒙古、朝鲜、日本、俄罗斯（远东）。
【生境】山地林下、林缘、灌丛。
【用途】药用。

Shortplume clematis
Buttercup family
Vine; leaves ternate or pinnate once to twice, leaflets with long-acuminate apex and notched margins, sometimes 3-lobed; paniculate cymes axillary or terminal; sepals 4, spreading, white or yellowish, densely pubescent along margins outer side; petals without; achenes numerous, persistent styles plumose. Flowering August to September.
Distribution: NE, N, NW, E and SW China; Mongolia, Korea, Japan, Russia (Far East)
Habitat: Montane woodlands, forest margins and scrublands
Use: Medicine

拍摄人：赵利清 Photo by Zhao Liqing

芍药
Paeonia lactiflora Pall.
芍药科 (牡丹科)
Paeoniaceae

【特征】多年生草本，高40~70cm；茎上部略分枝；叶为1至2回三出复叶，小叶狭卵形至披针形，边缘具齿；花数枚，顶生或腋生，有时仅顶端1枚开放；萼片3~4；花瓣9~13，白色或粉色；心皮3~5；蓇葖果具喙。花期5~7月。

【分布】东北、华北，陕西、甘肃；蒙古、朝鲜、日本、俄罗斯（西伯利亚、远东）。

【生境】山地和石质丘陵，林缘、灌丛、草甸和草原。

【用途】观赏；药用；提制栲胶。

Chinese peony
Peony family
Perennial herb 40~70 cm tall; stems slightly branched above; leaves ternate to biternate, leaflets narrowly ovate to lanceolate, serrate; several flowers terminal or axillary, sometimes only the terminal one developed; sepals 3~4; petals 9~13, white or pink; carpels 3~5; follicles beaked. Flowering May to July.

Distribution: NE and N China, Shaanxi and Gansu; Mongolia, Korea, Japan, Russia (Siberia and Far East)

Habitat: Mountains and rocky hills, forest margins, scrublands, meadows and steppe

Use: Ornamental, medicine; extracting tannin

贺兰山南芥
Arabis alaschanica **Maxim.**
十字花科
Brassicaceae

【特征】多年生矮小草本，高5～15cm；茎直立或倾斜；叶莲座状，肉质，叶柄具狭翅，叶片倒披针形至倒卵形，边缘有疏细齿和睫毛；花茎自基部抽出，具数花；花瓣4，白色至淡紫色；长角果狭条状，果梗劲直。花期5～6月。
【分布】内蒙古、宁夏、甘肃、四川。
【生境】海拔1 900～3 000m的山地草甸和山地石缝。
【用途】饲用；药用。

Alashan rockcress
Mustard family
Perennial dwarf herb 5～15 cm tall; stems erect or ascending; leaves rosulate, fleshy, petioles narrowly winged, leaf blades oblanceolate to obovate, margins spaced-serrulate and ciliate; flowering stems arising from base, with several flowers; petals 4, white to pale purple; siliques narrowly linear, fruiting pedicels straight. Flowering May to June.
Distribution: Inner Mongolia, Ningxia, Gansu and Sichuan
Habitat: Montane meadows and rock crevices at 1 900～3 000 m
Use: Forage; medicine

拍摄人：赵利清 Photo by Zhao Liqing

垂果南芥
Arabis pendula L.
十字花科
Brassicaceae

【特征】二年生草本，高30～150cm，全株被硬单毛，有时混生星状毛；茎直立，上部有分枝；叶矩圆状卵形至披针形，边缘具齿或近全缘，上部叶基部抱茎；总状花序顶生或腋生；花瓣4，白色；长角果条状，弧曲，下垂。花期6～8月。

【分布】东北、华北、西北、西南；亚洲北部和东部、欧洲。

【生境】海拔1 500～3 600m的林缘、灌丛、草甸。

【用途】饲用；药用。

Pendentfruit rockcress
Mustard family
Biennial herb 30～150 cm tall, simple-hispid throughout, sometimes mixed with stellate-hairs; stems erect, branched above; leaves oblong-ovate to lanceolate, serrate or subentire, upper leaves basally clasping; racemes terminal or axillary; petals 4, white; siliques linear, curved, drooping. Flowering June to August.
Distribution: NE, N, NW and SW China; N and E Asia, Europe
Habitat: Forest margins, thickets and meadows at 1 500～3 600 m
Use: Forage; medicine

串珠芥 (蚓果芥)
Neotorularia humilis (C. A. Mey.) Hedge et J. Léonard
(*Torularia humilis* (C. A. Mey.) O. E. Schulz)
十字花科
Brassicaceae

【特征】多年生草本，高5～30cm，全株被叉状毛或单毛；茎自基部分枝，直立、斜升或斜倚；叶全缘、浅波状或具疏齿；花序伞房状；花瓣4，白色或淡紫色；长角果条状，常呈念珠状，被叉状毛。花期5～9月。
【分布】华北、西北、四川、西藏；蒙古、朝鲜、西伯利亚、中亚、北美洲。
【生境】海拔1 000～4 200m的石质阳坡、石缝、沟谷。
【用途】饲用。

Low northern rockcress
Mustard family
Perennial herb 5～30 cm tall, forked-hairy or simple-hairy throughout; stems branched from base, erect, ascending or decumbent; leaves entire, repand or spaced-toothed; inflorescence corymbose; petals 4, white or pale purple; siliques linear, usually moniliform, forked-hairy. Flowering May to September.
Distribution: N and NW China, Sichuan and Tibet; Mongolia, Korea, Siberia, Central Asia, North America
Habitat: Sunny rocky slopes, rock crevices and ravines at 1 000～4 200 m
Use: Forage

东陵八仙花 (东陵绣球)
Hydrangea bretschneideri Dippel
虎耳草科
Saxifragaceae

- 【特征】灌木,高1~3m,当年枝红褐色至淡褐色;单叶对生,叶片卵形至长椭圆形,边缘具齿,下面被长柔毛;伞房状聚伞花序,花多数;不孕花萼片4,大型,卵形至近圆形,白色;孕花萼片小,长约1mm,三角形;花瓣5,白色。花期6~7月。
- 【分布】华北、陕西、宁夏、甘肃、青海、四川、湖北。
- 【生境】山地林缘、灌丛。
- 【用途】观赏;农具用材。

Shaggy hydrangea
Saxifrage family
Shrub 1~3 m tall; new branches red-brown to pale brown; simple leaves opposite, blades ovate to long-elliptic, toothed, pilose below; corymbose cymes with numerous flowers; sterile flowers with 4 large sepals ovate to suborbicular, white; fertile flowers with small sepals about 1 mm, triangular; petals 5, white. Flowering June to July.
Distribution: N China, Shaanxi, Ningxia, Gansu, Qinghai, Sichuan and Hubei
Habitat: Montane forest margins and scrublands
Use: Ornamental; materials for farm tools

拍摄人:赵利清 Photo by Zhao Liqing

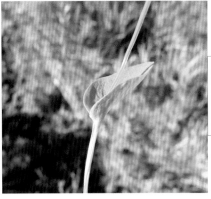

细叉梅花草
Parnassia oreophila Hance
虎耳草科
Saxifragaceae

【特征】多年生草本，高10～30cm，基生叶丛生，具长柄，花茎具1片无柄叶；花单生于花茎顶；花瓣5，白色；雄蕊5，退化雄蕊5，3裂；蒴果倒卵状。花期7～8月。

【分布】华北、陕西、宁夏、甘肃、青海、四川。

【生境】海拔1 600～3 000m的山地、沟谷、林下、林缘、草甸。

【用途】观赏；药用。

Mountain grass of parnassus
Saxifrage family
Perennial herb 10～30 cm tall; basal leaves tufted, long-petiolate; flowering stems with 1 sessile leaf; solitary flower terminal; petals 5, white; stamens 5, staminodia 5, 3-lobed; capsules obovoid. Flowering July to August.
Distribution: N China, Shaanxi, Ningxia, Gansu, Qinghai and Sichuan
Habitat: Mountains and ravines, forests, forest margins and meadows at 1 600～3 000 m
Use: Ornamental; medicine

堇叶山梅花
Philadelphus tenuifolius Rupr. ex Maxim.
虎耳草科
Saxifragaceae

【特征】灌木，高1.5～2m，当年枝紫褐色，光滑；单叶对生，叶片卵形至披针形，边缘具疏齿，上面绿色，下面灰绿色；总状花序具花5～9枚；萼片4，里面被短柔毛；花瓣4，乳白色；蒴果4瓣裂。花期6～7月。
【分布】东北、华北、陕西、甘肃、四川、河南、江苏、浙江；朝鲜。
【生境】山坡、林缘、灌丛。
【用途】观赏；药用。

Violetleaf mockorange
Saxifrage family
Shrub 1.5～2 m tall; new branches purple-brown, smooth; simple leaves opposite, blades ovate to lanceolate, spaced-toothed, green above, grey-green below; racemes with 5～9 flowers; sepals 4, pubescent inner side; petals 4, creamy-white; capsules 4-valved. Flowering June to July.
Distribution: NE and N China, Shaanxi, Gansu, Sichuan, Henan, Jiangsu and Zhejiang; Korea
Habitat: Slopes, forest edges and thickets
Use: Ornamental; medicine

拍摄人：赵利清 Photo by Zhao Liqing

拍摄人：赵利清 Photo by Zhao Liqing

水枸子
Cotoneaster multiflorus **Bunge**
蔷薇科
Rosaceae

【特征】灌木，高达2m；枝褐色或暗灰色；单叶互生，卵形、菱状卵形或椭圆形，全缘，两面无毛；聚伞花序腋生，花3～10枚；花梗和萼筒无毛；花瓣5，开展，白色，基部具柔毛；梨果近球状或宽卵状，鲜红色。花期6月。
【分布】东北、华北、西北、西南；西伯利亚、中亚、高加索。
【生境】山地、沟谷、林缘、灌丛。
【用途】水土保持；药用。

Many-flowered cotoneaster
Rose family
Shrub to 2 m tall; branches brown or dark grey; simple leaves alternate, ovate, rhombic-ovate or elliptic, entire, glabrous; cymes axillary, with 3～10 flowers; pedicel and calyx tube glabrous; petals 5, spreading, white, pubescent at base; pomes subglobose or broadly ovoid, bright-red. Flowering June.
Distribution: NE, N, NW and SW China; Siberia, Central Asia, Caucasia
Habitat: Mountains and ravines, forest edges and thickets
Use: Soil conservation; medicine

右图拍摄人：赵利清 Right photo by Zhao Liqing

准噶尔栒子
Cotoneaster soongoricus (Regel) Popov
蔷薇科
Rosaceae

【特征】灌木，高1～2.5m；枝灰褐色；单叶互生，卵形至椭圆形，全缘，下面被绒毛；聚伞花序，花3～5枚；花萼外面被绒毛；花瓣5，开展，白色；梨果卵状至椭圆状，红色，疏被绒毛。花期6～7月。
【分布】内蒙古西部、宁夏、甘肃、新疆、四川、西藏。
【生境】石质山坡。
【用途】水土保持。

Dzungar cotoneaster
Rose family
Shrub 1～2.5 m tall; branches grey-brown; simple leaves alternate, ovate to elliptic, entire, tomentose below; cymes with 3～5 flowers; calyx tomentose outer side; petals 5, spreading, white; pomes ovoid to ellipsoid, red, sparsely tomentose. Flowering June to July.
Distribution: W Inner Mongolia, Ningxia, Gansu, Xinjiang, Sichuan and Tibet
Habitat: Rocky slopes
Use: Soil conservation

蚊子草
Filipendula palmata (Pall.) Maxim.
蔷薇科
Rosaceae

【特征】多年生草本，高约1m；茎直立；单数羽状复叶，小叶5，顶生小叶5～9掌状深裂，侧生小叶3深裂，裂片边缘有齿，茎上部叶有小叶1～3，小叶片3～7掌状深裂，下面被毡毛；圆锥花序大型，花多数；花瓣5，白色。花期7月。
【分布】东北、华北；朝鲜、日本、俄罗斯（西伯利亚、远东）。
【生境】林下、林缘、灌丛、草甸。
【用途】观赏；药用；提取栲胶。

Siberian meadowsweet
Rose family
Perennial herb about 1 m tall; stems erect; leaves odd-pinnate, leaflets 5, terminal one palmately 5- to 9-parted, lateral ones 3-parted, segments serrate, upper leaves with leaflets 1～3, palmately 3- to 7-parted, manicate below; panicles large, with numerous flowers; petals 5, white. Flowering July.
Distribution: NE and N China; Korea, Japan, Russia (Siberia and Far East)
Habitat: Forests, forest margins, thickets and meadows
Use: Ornamental; medicine; extracting tannin

拍摄人：赵 凡 Photos by Zhao Fan

银露梅
Potentilla glabra Lodd.
蔷薇科
Rosaceae

【特征】灌木,高30～100cm;茎多分枝;单数羽状复叶,小叶3～5,近革质,全缘,边缘反卷,两面无毛或下面疏被柔毛;花单生或数枚;花瓣5,白色。花期6～8月。

【分布】华北、甘肃、青海、四川、云南、湖北、安徽;朝鲜、俄罗斯。

【生境】海拔1 000～4 500m的河谷、山地或高山灌丛。

【用途】饲用;药用;观赏;纤维材料;提制栲胶。

Snowpetal cinquefoil
Rose family
Shrub 30～100 cm tall; stems much branched; leaves odd-pinnate, leaflets 3～5, subleathery, entire, margins revolute, glabrous both sides or sparsely pubescent below; flowers solitary or several; petals 5, white. Flowering June to August.
Distribution: N China, Gansu, Qinghai, Sichuan, Yunnan, Hubei and Anhui; Korea, Russia
Habitat: River valleys and montane or alpine scrublands at 1 000～4 500 m
Use: Forage; medicine; ornamental; fiber material; extracting tannin

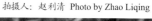
拍摄人: 赵利清 Photo by Zhao Liqing

拍摄人：赵利清 Photo by Zhao Liqing

华北覆盆子
Rubus idaeus var. *borealisinensis* T. T. Yu et L. T. Lu
蔷薇科
Rosaceae

【特征】灌木，高约1m；枝疏生皮刺；单数羽状复叶，小叶3～5，叶柄有皮刺，小叶片卵形或宽卵形，边缘有锯齿或重锯齿，下面密被绒毛；花序伞房状；总花梗、花梗和花萼密被短柔毛和腺毛，具针刺或无；花瓣5，白色；聚合果近球形，多汁液，红色。花期7～8月。
【分布】华北。
【生境】山地林缘、灌丛、草甸。
【用途】药用；果实可食用。

North China raspberry
Rose family
Shrub about 1 m tall; branches sparsely prickly; leaves odd-pinnate, leaflets 3～5, petioles prickly, leaflet blades ovate or broadly ovate, serrate or duplicato-serrate, densely tomentose below; inflorescence corymbose; peduncle, pedicel and calyx densely pubescent and glandular-hairy, with or without stiff bristles; petals 5, white; aggregate fruits subglobose, succulent, red. Flowering July to August.
Distribution: N China
Habitat: Montane forest margins, scrublands and meadows
Use: Medicine; fruits edible

库页悬钩子
Rubus sachalinensis H. Lév.
蔷薇科
Rosaceae

【特征】灌木，高0.6～2m；枝具皮刺；羽状三出复叶，叶柄被柔毛和疏刺，或混生腺毛，小叶片宽卵形至披针状卵形，边缘有锯齿，下面密被毡毛；花序伞房状；总花梗、花梗和花萼密被柔毛、腺毛和刺；花瓣5，白色；聚合果近球形，多汁液，红色。花期6～7月。
【分布】东北、华北、西北；蒙古、朝鲜、日本、俄罗斯、欧洲。
【生境】山地林下、林缘、灌丛、草甸。
【用途】药用；果实可食用。

Kuye raspberry (Sakhalin raspberry)
Rose family
Shrub 0.6～2 m tall; branches prickly; leaves pinnately trifoliolate, petioles pubescent and sparsely prickly or mixed with glandular hairs, leaflet blades broadly ovate to lanceolate-ovate, serrate, densely manicate below; inflorescence corymbose; peduncle, pedicel and calyx densely pubescent, glandular-hairy and bristly; petals 5, white; aggregate fruits subglobose, succulent, red. Flowering June to July.
Distribution: NE, N and NW China; Mongolia, Korea, Japan, Russia, Europe
Habitat: Montane forests, forest margins, scrublands and meadows
Use: Medicine; fruits edible

拍摄人：赵利清　Photos by Zhao Liqing

拍摄人：赵利清 Photos by Zhao Liqing

蒙古绣线菊
Spiraea mongolica Maxim.
蔷薇科
Rosaceae

【特征】灌木，高1~2m；枝褐色，具棱，无毛；叶片长椭圆形或椭圆状倒披针形，全缘，稀先端2~5浅裂，两面无毛；伞房花序具花8~17枚；花瓣5，白色；雄蕊与花瓣近等长；蓇葖果被短柔毛。花期6~7月。
【分布】华北、陕西、甘肃、青海、四川、西藏、河南。
【生境】石质或砾质山坡和沟谷。
【用途】水土保持；观赏；花药用。

Mongolian meadowsweet
Rose family
Shrub 1~2 m tall; branches brown, angled, glabrous; leaf blades long-elliptic or elliptic-oblanceolate, entire, rarely 2- to 5-lobed at apex, glabrous; corymbs with 8~17 flowers; petals 5, white; stamens subequaling petals; follicles pubescent. Flowering June to July.
Distribution: N China, Shaanxi, Gansu, Qinghai, Sichuan, Tibet and Henan
Habitat: Rocky or gravelly slopes and ravines
Use: Soil conservation; ornamental; flowers for medicine

毛枝蒙古绣线菊
Spiraea mongolica var. *tomentulosa* T. T. Yu
蔷薇科
Rosaceae

【特征】与蒙古绣线菊(*Spiraea mongolica*)的区别为：枝明显呈"之"字形，小枝被短柔毛。
【分布】内蒙古（贺兰山）、宁夏、甘肃、西藏。
【生境】石质或砾质山坡和沟谷。
【用途】水土保持；观赏。

Tomentulose meadowsweet
Rose family
Difference to *Spiraea mongolica*: Branches distinctly zigzagged, branchlets pubescent.
Distribution: Inner Mongolia (Helan Mountains), Ningxia, Gansu and Tibet
Habitat: Rocky or gravelly slopes and ravines
Use: Soil conservation; ornamental

拍摄人：赵利清 Photos by Zhao Liqing

卵果黄耆
Astragalus grubovii Sanchir
豆科
Fabaceae (Leguminosae)

【特征】多年生草本，高5～20cm，全株灰绿色，密被开展的丁字毛；茎无或短缩；单数羽状复叶，小叶9～29，小叶片椭圆形至倒卵形；总状花序密集于叶丛基部；蝶形花冠淡黄色或近白色，龙骨瓣略短于翼瓣；荚果矩圆状卵形，稍膨胀，密被白色长柔毛。花期5～6月。
【分布】内蒙古、陕西、宁夏、甘肃、新疆；蒙古。
【生境】草原至荒漠地带的砾石质或沙质地、干河谷、山麓、湖盆边缘。
【用途】饲用。

Ovoidfruit milkvetch
Pea family
Perennial herb 5～20 cm tall, grey-green, with dense and spreading medifixed hairs throughout; acaulescent or stems shortened; leaves odd-pinnate, leaflets 9～29, blades elliptic to obovate; racemes congested in base of leaf cluster; papilionaceous corolla pale yellow or nearly white, keel slightly shorter than wings; pods oblong-ovoid, slightly inflated, densely white-villous. Flowering May to June.
Distribution: Inner Mongolia, Shaanxi, Ningxia, Gansu and Xinjiang; Mongolia
Habitat: Gravelly and sandy sites, dry valleys, foothills and lake basin edges in steppe to desert zones
Use: Forage

圆果黄耆 (尤那托夫黄耆)
Astragalus junatovii Sanchir
豆科
Fabaceae (Leguminosae)

- 【特征】多年生草本，高5～15cm；地上茎无或极短缩；单数羽状复叶，小叶5～15，小叶片椭圆形或披针形，两面密被白色伏贴毛；总状花序近无梗，具花2～4枚；蝶形花冠白色，龙骨瓣略短于翼瓣；荚果近球状，密被白绵毛。花期5～6月。
- 【分布】内蒙古西部、宁夏、甘肃；蒙古。
- 【生境】荒漠草原沙砾质地。
- 【用途】饲用。

Roundfruit milkvetch
Pea family
Perennial herb 5～15 cm tall; acaulescent or stems strongly shortened; leaves odd-pinnate, leaflets 5～15, blades elliptic or lanceolate, densely white appressed-hairy; racemes subsessile, with 2～4 flowers; papilionaceous corolla white, keel slightly shorter than wings; pods subglobose, densely white-lanate. Flowering May to June.
Distribution: W Inner Mongolia, Ningxia and Gansu; Mongolia
Habitat: Sandy and gravelly sites in desert-steppe
Use: Forage

拍摄人：赵利清 Photos by Zhao Liqing

拍摄人：赵利清 Photo by Zhao Liqing

短龙骨黄耆
Astragalus parvicarinatus S. B. Ho
豆科
Fabaceae (Leguminosae)

【特征】多年生草本，高5～10cm；地上茎无或极短缩；单数羽状复叶，小叶3～7，小叶片椭圆形至倒卵形，两面密被白色伏贴毛，呈灰绿色；花腋生，无梗；蝶形花冠白色，龙骨瓣长约为翼瓣的1/2。花期4～5月。
【分布】内蒙古西部、宁夏。
【生境】荒漠地带的沙砾质地。
【用途】饲用。

Shortkeel milkvetch
Pea family
Perennial herb 5～10 cm tall; acaulescent or stems strongly shortened; leaves odd-pinnate, leaflets 3～7, blades elliptic or obovate, densely white appressed-hairy, grey-green; flowers axillary, sessile; papilionaceous corolla white, keel 1/2 as long as wings. Flowering April to May.
Distribution: W Inner Mongolia and Ningxia
Habitat: Sandy and gravelly sites in desert zone
Use: Forage

苦参
Sophora flavescens Aiton
豆科
Fabaceae (Leguminosae)

【特征】多年生草本或半灌木，高1～2m；茎直立，多分枝；单数羽状复叶，小叶11～19，小叶片卵状矩圆形至狭卵形；总状花序顶生，花多数；蝶形花冠白色或黄白色；荚果近念珠状，长5～12cm。花期6～7月。
【分布】东北、华北；朝鲜、日本、俄罗斯（远东）。
【生境】草原带沙地、山坡、田埂。
【用途】根药用；种子可作农药；茎皮为纤维材料。

Ku shen (Yellow sophora)
Pea Family
Perennial or subshrub 1～2 m tall; stems erect, much branched; leaves odd-pinnate, leaflets 11～19, blades ovate-oblong to narrowly ovate; racemes terminal, with numerous flowers; papilionaceous corolla white or yellowish-white; pods nearly torulose, 5～12 cm long. Flowering June to July.
Distribution: NE and N China; Korea, Japan, Russia (Far East)
Habitat: Sands, slopes and field ridges in steppe zone
Use: Roots for medicine; seeds for making pesticide; stem coats for fiber materials

拍摄人：赵利清 Photo by Zhao Liqing

短喙牻牛儿苗 (藏牻牛儿苗)
Erodium tibetanum Edgew.
牻牛儿苗科
Geraniaceae

【特征】一年生或二年生矮小草本，高2～6cm；无茎；叶基生，莲座状，叶片1至2回羽状分裂，两面被毡毛，灰绿色；花2～4枚；萼片被毡毛，先端稍钝或圆；花瓣5，通常白色；蒴果具喙，喙长12～14mm。花期5～7月。
【分布】内蒙古西部、宁夏、甘肃、新疆、西藏；蒙古、塔吉克斯坦、巴基斯坦。
【生境】荒漠草原至荒漠地带的戈壁、沙丘、干河床。
【用途】饲用。

Tibetan stork's-bill
Geranium family
Annual or biennial dwarf herb 2～6 cm tall; acaulescent; leaves basal, rosulate, leaf blades pinnately parted once to twice, manicate, grey-green; flowers 2～4; sepals manicate, apex slightly obtuse or rounded; petals 5, usually white; capsules with a beak to 12～14 mm long. Flowering May to July.
Distribution: W Inner Mongolia, Ningxia, Gansu, Xinjiang and Tibet; Mongolia, Tajikistan, Pakistan
Habitat: Gobi, dunes and dry riverbeds in desert-steppe to desert zones
Use: Forage

骆驼蓬
Peganum harmala L.
蒺藜科
Zygophyllaceae

- 【特征】多年生草本，高30～80cm，全株无毛；茎直立或开展，由基部多分枝；叶3～5全裂，裂片条形或披针状条形；花单生；花瓣5，黄白色。花期5～6月。
- 【分布】内蒙古西部、宁夏、甘肃、新疆；蒙古、俄罗斯、巴基斯坦、阿富汗、中亚、西亚、北非、南欧。
- 【生境】荒漠地带的干旱草地、轻度盐渍地、村旁、路边。
- 【用途】全草药用并制杀虫剂；种子做红色染料并提取工业用油。

Harmel peganum (Syrian rue)
Caltrop family
Perennial herb 30～80 cm tall, glabrous throughout; stems erect or spreading, much branched from base; leaves 3- to 5-divided, segments linear to lanceolate-linear; flowers solitary; petals 5, yellowish-white. Flowering May to June.
Distribution: W Inner Mongolia, Ningxia, Gansu and Xinjiang; Mongolia, Russia, Pakistan, Afghanistan, Central Asia, W Asia, N Africa, S Europe
Habitat: Dry grasslands, lightly saline sites, village sides and roadsides in desert zone
Use: Whole plant for medicine and for making insecticide;
seeds for making red dyestuff and extracting industrial oil

拍摄人：赵利清 Photo by Zhao Liqing

鸡腿堇菜
Viola acuminata Ledeb.
堇菜科
Violaceae

【特征】多年生草本，高5～50cm；茎直立，丛生；托叶大，羽状深裂；叶片心状卵形或卵形，边缘具齿；花两侧对称；花瓣5，白色或淡紫色，最下瓣片具紫色脉纹，基部具距，距长3～4mm。花果期5～9月。
【分布】东北、华北、华中；朝鲜、日本、俄罗斯（西伯利亚、远东）。
【生境】疏林、林缘、灌丛、草甸。
【用途】饲用；药用。

Drumstick violet
Violet family
Perennial herb 5～50 cm tall; stems erect, tufted; stipule large, pinnatipartite; leaf blades cordate-ovate or ovate, serrate; flowers zygomorphic; petals 5, white or pale purple, the lower one with purple streaks and with a basal spur 3～4 mm long. Flowering and fruiting May to September.
Distribution: NE, N and C China; Korea, Japan, Russia (Siberia and Far East)
Habitat: Open woodlands, forest edges, thickets and meadows
Use: Forage; medicine

短茎古当归
Archangelica brevicaulis (Rupr.) Rchb.
伞形科
Apiaceae (Umbelliferae)

【特征】多年生草本，高40～100cm；茎直立，中空；叶2～3回羽状分裂，顶生末回裂片不分裂，叶片两面被毛，茎生叶叶柄下部膨大成阔囊状叶鞘；复伞形花序，伞幅20～50；花瓣5，白色；果棱厚翅状；棱槽内油管3～4，合生面油管6～7。花期7～8月。
【分布】新疆西部；中亚。
【生境】河谷草甸。
【用途】药用；饲用。

Shortstem archangelica
Parsley family
Perennial herb 40～100 cm tall; stems erect, hollow; leaves pinnately cleft twice to thrice, the terminal-ultimate segment not lobed, blades pubescent, petioles of cauline leaf basally inflated into a broadly saccate sheath; compound umbels with 20～50 rays; petals 5, white; schizocarp ribs thickly wing-like; oil tubes 3～4 in the intervals, 6～7 on the commissure. Flowering July to August.
Distribution: W Xinjiang; Central Asia
Habitat: Meadows in river valleys
Use: Medicine; forage

拍摄人：张洪江 Photo by Zhang Hongjiang

拍摄人：张洪江 Photos by Zhang Hongjiang

下延叶古当归
Archangelica decurrens Ledeb.
伞形科
Apiaceae (Umbelliferae)

【特征】多年生草本，高1~2m；茎直立，中空；叶三出式2~3回羽状全裂，顶生末回裂片常3浅裂，叶片两面无毛，茎生叶叶柄下部膨大成囊状叶鞘；复伞形花序，伞幅20~50；花瓣5，白色；果棱厚翅状；油管极多，连成环状。花期7~8月。

【分布】新疆、内蒙古；蒙古、俄罗斯。

【生境】山谷、林下、沟边。

【用途】药用；饲用。

Decurrent archangelica
Parsley family
Perennial herb 1~2 m tall; stems erect, hollow; leaves ternately pinnatisect twice to thrice, the terminal-ultimate segment usually 3-lobed, glabrous, petioles of cauline leaf basally inflated into a saccate sheath; compound umbels with 20~50 rays; petals 5, white; schizocarp ribs thickly wing-like; oil tubes numerous, connected into a ring. Flowering July to August.
Distribution: Xinjiang and Inner Mongolia; Mongolia, Russia
Habitat: Valleys, forests and gully sides
Use: Medicine; forage

田葛缕子
Carum buriaticum Turcz.
伞形科
Apiaceae (Umbelliferae)

【特征】二年生草本，高25～80cm；茎直立，自下部分枝；叶3～4回羽状全裂，末回裂片条形，茎生叶叶鞘边缘白色狭膜质；复伞形花序，伞幅8～12；小总苞片5～8；花瓣5，白色。花期7～8月。
【分布】东北、华北、西北、四川、西藏；蒙古、俄罗斯（西伯利亚、远东）。
【生境】草甸、撂荒地、田边、沟谷、路旁。
【用途】饲用；药用；制作食用香料；提取挥发油。

Field caraway
Parsley family
Biennial herb 25～80 cm tall; stems erect, branched from the lower portion; leaves pinnatisect 3 to 4 times, ultimate segments linear, sheaths of cauline leaf with white and narrowly membranous margins; compound umbels with 8～12 rays; bracteoles 5～8; petals 5, white. Flowering July to August.
Distribution: NE, N and NW China, Sichuan and Tibet; Mongolia, Russia (Siberia and Far East)
Habitat: Meadows, abandoned lands, farmland sides, gullies and roadsides
Use: Forage; medicine; making edible spice; extracting volatile oil

拍摄人：赵 凡 Photos by Zhao Fan

碱蛇床
Cnidium salinum Turcz.
伞形科
Apiaceae (Umbelliferae)

【特征】二年生或多年生草本，高20～50cm；茎直立或下部稍膝曲，上部稍分枝；叶2～3回羽状全裂，末回裂片条形，边缘稍卷折；复伞形花序，伞幅8～15；小总苞片3～6，长于花梗；花瓣5，白色；分果具棱翅。花期8月。
【分布】黑龙江、内蒙古、宁夏、甘肃、青海；蒙古、西伯利亚。
【生境】草甸、盐碱滩、河边、沟渠旁。
【用途】饲料；提取芳香油。

Alkali snowparsley
Parsley family
Biennial or perennial herb 20～50 cm tall; stems erect or slightly geniculate at lower portion, slightly branched above; leaves pinnatisect twice to thrice, ultimate segments linear, margins slightly involute-folded; compound umbels with 8～15 rays; bracteoles 3～6, longer than the pedicels; petals 5, white; schizocarps with winged ribs. Flowering August.
Distribution: Heilongjiang, Inner Mongolia, Ningxia, Gansu and Qinghai; Mongolia, Siberia
Habitat: Meadows, saline-alkali sites, riversides and ditch sides
Use: Forage; extracting essential oil

柳叶芹
Czernaevia laevigata **Turcz.**
伞形科
Apiaceae (Umbelliferae)

【特征】二年生草本，高40～100cm；茎单一，直立；叶2回羽状全裂，末回裂片披针形至矩圆状披针形，边缘具白色软骨质锯齿或重锯齿；复伞形花序，伞幅15～30；小总苞片2～8，短于花梗；花瓣5，白色；分果具棱翅。花期7～8月。

【分布】东北、华北；朝鲜、俄罗斯（西伯利亚、远东）。

【生境】山地灌丛、林下、林缘、草甸、沼泽草甸。

【用途】饲用；幼苗叶可食。

Willowleaf celery
Parsley family
Biennial herb 40～100 cm tall; stems single, erect; leaves bipinnatisect, ultimate segments lanceolate to oblong-lanceolate, margins serrate or duplicato-serrate with white cartilaginous tip; compound umbels with 15～30 rays; bracteoles 2～8, shorter than the pedicels; petals 5, white; schizocarps with winged ribs. Flowering July to August.
Distribution: NE and N China; Korea, Russia (Siberia and Far East)
Habitat: Montane scrublands, forests, forest borders, meadows and swamp-meadows
Use: Forage; seedling leaves edible

拍摄人：赵 凡 Photos by Zhao Fan

短毛独活
Heracleum moellendorffii Hance
伞形科
Apiaceae (Umbelliferae)

【特征】多年生草本，高80～200cm，全株被毛；茎直立，上部分枝；叶羽状复叶或2回羽状分裂，侧生小叶3～5浅裂或深裂，裂片边缘有粗齿，茎上部叶叶鞘膨大；复伞形花序顶生和侧生，伞幅12～30；花瓣5，白色，外缘花的外侧花瓣增大，2裂；果实侧棱宽翅状。花期7～8月。
【分布】东北、华北、西北、西南、华东；亚洲北部、北美洲。
【生境】林下、林缘、田边、山沟溪边。
【用途】药用。

Wooly cowparsnip
Parsley family
Perennial herb 80～200 cm tall, hairy throughout; stems erect, branched above; leaves pinnate or bipinnatifid, lateral leaflets 3- to 5-lobed or parted, segments coarsely serrate, upper leaves with inflated sheaths; compound umbels terminal and lateral, with 12～30 rays; petals 5, white, outer petals of the peripheral flowers enlarged and 2-lobed; lateral ribs of fruit broadly wing-like. Flowering July to August.
Distribution: NE, N, NW, SW and E China; N Asia, North America
Habitat: Forests, forest margins, farmland sides and gully streamsides
Use: Medicine

绿花山芹
Ostericum viridiflorum (Turcz.) Kitag.
伞形科
Apiaceae (Umbelliferae)

【特征】二年生或多年生草本，高50~100cm；茎直立，有分枝；叶2回三出羽状复叶，小叶片卵形或披针状卵形，边缘具大齿；复伞形花序顶生和侧生，侧生者花序梗较长，伞幅10~18；小总苞片5~9；花瓣5，淡绿色或白色，基部具长爪；分果侧棱具宽翅。花期7~8月。

【分布】东北，内蒙古东部；俄罗斯（西伯利亚、远东）。

【生境】林缘、草甸和沼泽草甸。

【用途】饲用；幼苗可食用；果实提取芳香油。

Green hillcelery
Parsley family
Biennial or perennial herb 50~100 cm tall; stems erect, branched; leaves biternately pinnate, leaflet blades ovate or lanceolate-ovate, large-serrate; compound umbels terminal and lateral, the lateral ones with longer peduncle, rays 10~18; bracteoles 5~9; petals 5, pale green or white, basally long-clawed; lateral ribs of schizocarps broadly winged. Flowering July to August.
Distribution: NE China, E Inner Mongolia; Russia (Siberia and Far East)
Habitat: Forest margins, meadows and swamp-meadows
Use: Forage; seedlings edible; fruits for extracting essential oil

华北前胡
Peucedanum harry-smithii Fedde ex Wolff
(毛白花前胡 *Peucedanum praeruptorum* ssp. *hirsutiusculum* Ma)
伞形科
Apiaceae (Umbelliferae)

【特征】多年生草本，高40～100cm；茎直立，上部稍分枝；叶2～3回羽状全裂，末回裂片近菱形或卵状披针形，边缘具少数缺刻状牙齿，下面灰蓝色；复伞形花序，伞幅10～20；小总苞片6～9，边缘宽膜质；花瓣5，白色；分果密被极细短硬毛，侧棱具狭翅。花期7～8月。
【分布】华北，山东、河南、陕西、甘肃。
【生境】山坡林缘、山沟溪边。
【用途】饲用；药用。

Wooly hogfennel
Parsley family
Perennial herb 40～100 cm tall; stems erect, slightly branched above; leaves pinnatisect twice to thrice, ultimate segments subrhombic or ovate-lanceolate, margins with few incised teeth, grey-blue below; compound umbels with 10～20 rays; bracteoles 6～9, with widely membranous margins; petals 5, white; schizocarps densely minute-puberulent, lateral ribs narrowly winged. Flowering July to August.
Distribution: N China, Shandong, Henan, Shaanxi and Gansu
Habitat: Sloping forest margins and gully streamsides
Use: Forage; medicine

棱子芹
Pleurospermum uralense Hoffm.
伞形科
Apiaceae (Umbelliferae)

【特征】多年生草本，高70～150cm；茎直立；叶2～3回羽状全裂，末回裂片狭卵形至披针形，边缘羽状缺刻或具齿；复伞形花序顶生和侧生，侧生者长于顶生者，伞幅20～40；总苞片多数，向下反折，羽状深裂；小总苞片10余片，条形，向下反折；花瓣5，白色，无小舌片。花期6～7月。

【分布】东北、华北；蒙古、朝鲜、日本、俄罗斯（远东）。

【生境】山谷林下、林缘、草甸、溪边。

【用途】饲用；药用。

Ural ribseedcelery
Parsley family
Perennial herb 70～150 cm tall; stems erect; leaves pinnatisect twice to thrice, ultimate segments narrowly ovate to lanceolate, margins pinnately incised or toothed; compound umbels terminal and lateral, the lateral ones longer than the terminal one, rays 20～40; bracts numerous, reflexed, pinnatipartite; bracteoles more than 10, linear, reflexed; petals 5, white, ligule without. Flowering June to July.
Distribution: NE and N China; Mongolia, Korea, Japan, Russia (Far East)
Habitat: Valley forests, forest margins, meadows and streamsides
Use: Forage; medicine

拍摄人：赵 凡（左图）& 赵利清（右图）Photos by Zhao Fan & Zhao Liqing

红瑞木
Swida alba Opiz
(*Cornus alba* L.)
山茱萸科
Cornaceae

【特征】灌木，高达2m；小枝紫红色，光滑；单叶对生，叶片卵状椭圆形或宽卵形，全缘，下面疏生长柔毛；伞房状聚伞花序顶生，花多数；花瓣4，白色或黄白色；核果矩圆状或近球状，乳白色；核扁平。花期5～6月。

【分布】东北、内蒙古、河北、陕西、山东、江苏；蒙古、朝鲜。

【生境】河岸、溪边。

【用途】庭院绿化；种子可提炼工业用油。

Tatarian dogwood
Dogwood family
Shrub to 2 m tall; branchlets purple-red, smooth; simple leaves opposite, blades ovate-elliptic or widely ovate, entire, sparsely villous beneath; corymbose cymes terminal, with numerous flowers; petals 4, white or yellowish-white; drupes oblong or subglobose, creamy-white; pits flat. Flowering May to June.
Distribution: NE China, Inner Mongolia, Hebei, Shaanxi, Shandong and Jiangsu; Mongolia, Korea
Habitat: River banks and streamsides
Use: Courtyard planting; seeds for extracting industrial oil

拍摄人：赵利清 Photo by Zhao Liqing

沙梾
Swida bretschneideri (L. Henry) Sojak
(*Cornus bretschneideri* L. Henry)
山茱萸科
Cornaceae

【特征】灌木，高达2m；小枝紫红色或暗紫色，被短柔毛；单叶对生，叶片椭圆形至卵形，全缘，下面密被短毛；圆锥状聚伞花序顶生，花多数；花瓣4，白色；核果球状，蓝黑色；核球状卵形。花期5～6月。

【分布】东北、华北。

【生境】海拔1 500～2 300m的阴坡杂木林或灌丛。

【用途】庭院绿化。

Sandy dogwood
Dogwood family
Shrub to 2 m tall; branchlets purple-red or dark purple, pubescent; simple leaves opposite, blades elliptic to ovate, entire, densely short-hairy beneath; paniculate cymes terminal, with numerous flowers; petals 4, white; drupes globose, blue-black; pits globose-ovoid. Flowering May to June.

Distribution: NE and N China
Habitat: Shady sloping mixed-woodlands or scrublands at 1 500～2 300 m
Use: Courtyard planting

阿拉善点地梅
Androsace alaschanica Maxim.
报春花科
Primulaceae

【特征】多年生半灌木状草本，高2.5～4cm；茎多次叉状分枝，形成垫状密丛；枯叶基部宿存，鳞片状重叠覆盖于分枝上；新叶灰绿色，革质，条状披针形或近钻形；花葶单一，短，藏于叶丛中，顶生1～2花；花冠白色，裂片5。花期4～5月。

【分布】内蒙古西部、宁夏、甘肃、青海。

【生境】海拔1 500～2 200m的山地草原、石质坡地和沙地。

【用途】观赏。

Alashan rockjasmine
Primula family
Perennial suffrutescent herb 2.5～4 cm tall; stems forked branching several times, forming a dense cushion; withered leaves basally persistent, scale-like and superimposed on branches; fresh leaves grey-green, leathery, linear-lanceolate or nearly subulate; scape solitary, short, hiding in leaf clusters, with terminal flowers 1 or 2; corolla white, with 5 lobes. Flowering April to May.
Distribution: W Inner Mongolia, Ningxia, Gansu and Qinghai
Habitat: Montane steppe, rocky slopes and sands at 1 500～2 200 m
Use: Ornamental

拍摄人：赵利清 Photos by Zhao Liqing

小点地梅
Androsace gmelinii (L.) Roem. et Schult.
报春花科
Primulaceae

【特征】一年生矮小草本，全株被长柔毛；叶基生，具叶柄，叶片近圆形至圆肾形，边缘具圆齿；花葶数枚，高3～6cm；伞形花序具花2～5枚；花冠小，白色，裂片5。花期6～8月。
【分布】内蒙古、甘肃、青海、四川、新疆；蒙古、俄罗斯（西伯利亚、远东）。
【生境】山地、沟谷、河岸湿地、林缘草甸。
【用途】饲用。

Little rockjasmine
Primula family
Annual dwarf herb, villous throughout; leaves basal, petioled, blades suborbicular to rounded-reniform, crenate; scapes several, 3～6 cm tall; umbels with flowers 2～5; small corolla white, with 5 lobes. Flowering June to August.
Distribution: Inner Mongolia, Gansu, Qinghai, Sichuan and Xinjiang; Mongolia, Russia (Siberia and Far East)
Habitat: Mountains and ravines, wet sites along river banks and meadows around forest borders
Use: Forage

拍摄人：赵利清 Photo by Zhao Liqing

拍摄人：赵利清 Photo by Zhao Liqing

歧伞獐牙菜
Swertia dichotoma L.
龙胆科
Gentianaceae

【特征】一年生草本，高5~20cm；茎纤弱，斜升，具四棱，沿棱具狭翅，自基部二歧式分枝；基生叶匙形，茎生叶卵形或卵状披针形；聚伞花序通常3花，或单花；花梗细弱，弯垂；花冠辐状，白色或带紫色，4裂，裂片基部有2腺洼。花期7~8月。

【分布】东北、华北、西北、华中；蒙古、日本、西伯利亚、中亚。

【生境】山坡、河边、林缘、草甸。

【用途】观赏。

Forked felwort
Gentian family
Annual herb 5~20 cm tall; stems slim and fragile, ascending, quadrangular, narrowly winged on acies, dichotomous from base; basal leaves spatulate, cauline leaves ovate or ovate-lanceolate; flowers usually 3 in a cyme, or solitary; pedicels slender, nodding; corolla rotate, white or purplish, 4-lobed, with 2 nectaries at lobe base. Flowering July to August.
Distribution: NE, N, NW and C China; Mongolia, Japan, Siberia, Central Asia
Habitat: Slopes, riversides, forest margins and meadows
Use: Ornamental

椭圆叶天芥菜
Heliotropium ellipticum **Ledeb.**
紫草科
Boraginaceae

【特征】多年生草本，高20～50cm，全株密被糙硬毛；茎直立或斜升，自基部分枝；叶椭圆形或椭圆状卵形；聚伞花序顶生，单一或2叉分枝；花冠白色，辐状，5裂。花期7～9月。

【分布】新疆；中亚、巴基斯坦、伊朗。

【生境】山地草坡、沟谷、路旁。

【用途】饲用。

Elliptic heliotrope
Borage family
Perennial herb 20～50 cm tall, densely hispid throughout; stems erect or ascending, branched from base; leaves elliptic or elliptic-ovate; cymes terminal, single or bifurcately branched; corolla white, rotate, 5-lobed. Flowering July to September.

Distribution: Xinjiang; Central Asia, Pakistan, Iran
Habitat: Mountain grassy slopes, ravines and roadsides
Use: Forage

北方拉拉藤
Galium boreale L.
茜草科
Rubiaceae

【特征】多年生草本，高15～65cm；茎直立，具四棱；叶4片轮生，披针形至狭披针形，基出脉3条，两面无毛；聚伞状圆锥花序顶生，花密集；萼筒密被钩状毛；花冠白色，辐状，裂片4；果实密被钩状毛。花期7月。
【分布】东北、华北、西北；朝鲜、日本、西伯利亚、北欧、北美洲。
【生境】山地林下、林缘、灌丛、草甸。
【用途】饲用；药用。

Northern bedstraw
Madder family
Perennial herb 15～65 cm tall; stems erect, quadrangular; leaves in whorls of 4, lanceolate to narrowly lanceolate, 3-basinerved, glabrous both sides; thyrse terminal, with dense flowers; calyx tube densely glochidiate; corolla white, rotate, with 4 lobes; fruits densely glochidiate. Flowering July.
Distribution: NE, N and NW China; Korea, Japan, Siberia, N Europe, North America
Habitat: Montane forests, forest edges, thickets and meadows
Use: Forage; medicine

茜草
Rubia cordifolia L.
茜草科
Rubiaceae

【特征】多年生草本；茎蔓生，小枝四棱形，棱上具倒生小刺；叶4～8片轮生，卵状披针形或卵形，边缘具倒生小刺，下面疏被刺状毛，脉上有倒生小刺；聚伞状圆锥花序顶生或腋生；花冠黄白色，辐状，裂片5；果实橙红色，熟时不变黑。花期7月。
【分布】中国大部分地区；亚洲北部至澳大利亚。
【生境】林下、林缘、灌丛、草甸、路边。
【用途】根药用并可做染料。

Indian madder
Madder family
Perennial herb; stems trailing, branchlets quadrangular, barbellulate on acies; leaves in whorls of 4～8, ovate-lanceolate or ovate, with barbellulate edges, sparsely setaceous-hairy below, barbellulate on veins; thyrse terminal or axillary; corolla yellow-white, rotate, with 5 lobes; fruits orange-red, never black even ripe. Flowering July.
Distribution: Most regions of China; N Asia to Australia
Habitat: Forests, forest edges, thickets, meadows and roadsides
Use: Roots for medicine and for making dyestuff

右图拍摄人：赵利清 Right photo by Zhao Liqing

小叶忍冬
Lonicera microphylla **Willd. ex Roem. et Schult.**
忍冬科
Caprifoliaceae

【特征】灌木，高1~1.5m；小枝灰褐色；叶倒卵形至矩圆形，边缘具睫毛，两面密被柔毛，有时光滑；花冠黄白色，二唇形，花冠筒基部浅囊状，上唇4浅裂，边缘具毛，下唇具1裂片；浆果橙红色，球状。花期5~6月。
【分布】西北、内蒙古；蒙古、中亚。
【生境】山地、丘陵、石崖，疏林、灌丛。
【用途】水土保持；庭院绿化。

Littleleaf honeysuckle
Honeysuckle family
Shrub 1~1.5 m tall; branchlets grey-brown; leaves obovate to oblong, fringed at margins, densely pubescent or sometimes glabrous; corolla yellowish-white, bilabiate, tube shallowly saccate at base, upper lip 4-lobed with ciliate edges, lower lip with 1 lobe; berries orangered, globose. Flowering May to June.
Distribution: NW China, Inner Mongolia; Mongolia, Central Asia
Habitat: Mountains, hills and rock precipices, open woods and thickets
Use: Soil conservation; courtyard planting

大青山风铃草
Campanula glomerata ssp. *daqingshanica* Hong et Y. Z. Zhao
桔梗科
Campanulaceae

【特征】多年生草本，高14～50cm；茎直立，单一，少分枝；基生叶长椭圆形或卵状披针形，基部圆形，茎上部叶卵状三角形，基部半抱茎；花簇生茎顶，少腋生；花冠蓝紫色或白色，筒状钟形，5中裂。花期7～8月。

【分布】内蒙古。

【生境】山地草甸、林缘、林间草甸。

【用途】饲用；药用。

Daqingshan bellflower
Bellflower family
Perennial herb 14～50 cm tall; stems erect, single, rarely branched; basal leaves long-elliptic or ovate-lanceolate with rounded base, upper leaves ovate-triangular, base clasping; clustered flowers terminal, rarely axillary; corolla blue-purple or white, tubular-campanulate, 5-cleft. Flowering July to August.
Distribution: Inner Mongolia
Habitat: Montane meadows, forest margins, meadows in forest openings
Use: Forage; medicine

拍摄人：赵利清 Photos by Zhao Liqing

拍摄人：赵 凡 Photos by Zhao Fan

紫斑风铃草
Campanula punctata Lam.
桔梗科
Campanulaceae

【特征】多年生草本，高20～50cm，全株被刚毛；茎直立；基生叶心状卵形，茎生叶三角状卵形至披针形，边缘具钝齿；花单生于茎顶或叶腋，下垂；花冠白色，带紫斑，筒状钟形，5浅裂。花期6～8月。
【分布】东北、华北、华中；朝鲜、日本、俄罗斯（远东）。
【生境】草甸、林缘、灌丛。
【用途】观赏；药用。

Spotted bellflower
Bellflower family
Perennial herb 20～50 cm tall, setose throughout; stems erect; basal leaves cordate-ovate, cauline leaves triangular-ovate to lanceolate, margins obtuse-toothed; solitary flower terminal or axillary, nodding; corolla white with purple spots, tubular-campanulate, 5-lobed. Flowering June to August.
Distribution: NE, N and C China; Korea, Japan, Russia (Far East)
Habitat: Meadows, forest edges and thickets
Use: Ornamental; medicine

高山蓍
Achillea alpine L.
菊科
Asteraceae (Compositae)

【特征】多年生草本，高30～70cm；茎直立，上部有分枝；叶羽状浅裂至深裂，裂片有缺刻状锯齿，两面疏被长柔毛；头状花序多数，密集成伞房状；舌状小花7～8，白色，管状小花5齿裂，白色。花期7～8月。

【分布】东北、华北，宁夏、甘肃；蒙古、朝鲜、日本、俄罗斯（西伯利亚、远东）、中亚。

【生境】山地林缘、灌丛、草甸。

【用途】饲用；药用。

Alpine yarrow
Aster family
Perennial herb 30～70 cm tall; stems erect, branched above; leaves pinnately lobed to parted, segments incised-serrate, sparsely villous; numerous heads in a corymbiform cluster; ray florets 7～8, white, disk florets 5-toothed, white. Flowering July to August.
Distribution: NE and N China, Ningxia and Gansu; Mongolia, Korea, Japan, Russia (Siberia and Far East), Central Asia
Habitat: Montane forest margins, thickets and meadows
Use: Forage; medicine

拍摄人：赵 凡 Photos by Zhao Fan

细叶菊
Dendranthema maximowiczii **(Komar) Tzvelev**
菊科
Asteraceae (Compositae)

【特征】二年生草本，高15～30cm；茎直立，单生，中上部有少数分枝；叶2回羽状全裂或几全裂，小裂片条形；头状花序2～4，排列成伞房状，稀单生；舌状小花白色或粉红色，管状小花黄色。花期7～8月。
【分布】东北，内蒙古东部；朝鲜、俄罗斯（远东）。
【生境】山坡灌丛、草甸。
【用途】饲用。

Thinleaf daisy
Aster family
Biennial herb 15～30 cm tall; stems erect, single, with a few branches in upper half; leaves bipinnatisect or nearly so, lobules linear; heads 2～4, corymbiform-arranged, rarely solitary; ray florets white or pink, disk florets yellow. Flowering July to August.
Distribution: NE China, E Inner Mongolia; Korea, Russia (Far East)
Habitat: Sloping thickets and meadows
Use: Forage

楔叶菊
Dendranthema naktongense (Nakai) Tzvelev
菊科
Asteraceae (Compositae)

【特征】多年生草本，高15~50cm；茎直立；叶掌式羽状或羽状浅裂至深裂，叶片基部楔形；头状花序2~9，排列成伞房状，稀单生；舌状小花白色、粉红色或淡紫色，管状小花黄色。花期7~8月。
【分布】东北，内蒙古、河北；朝鲜、日本、俄罗斯（远东）。
【生境】林缘、灌丛、草甸。
【用途】饲用。

Wedgeleaf daisy
Aster family
Perennial herb 15~50 cm tall; stems erect; leaves palmate-pinnately or pinnately lobed to parted, base cuneate; heads 2~9, corymbiform-arranged, rarely solitary; ray florets white, pink or pale purple, disk florets yellow. Flowering July to August.
Distribution: NE China, Inner Mongolia and Hebei; Korea, Japan, Russia (Far East)
Habitat: Forest edges, thickets and meadows
Use: Forage

革苞菊
Tugarinovia mongolica Iljin
菊科
Asteraceae (Compositae)

【特征】多年生草本，植株有胶黏液汁；茎基被厚绵毛，花茎长2～4cm；叶基生，莲座状，革质，羽状深裂至浅裂，裂片皱曲，具浅齿和硬刺，两面被蛛丝状毛；头状花序单生于茎顶，雄花序较小；外层总苞片革质；管状小花白色。花期5～6月。
【分布】内蒙古中部；蒙古。
【生境】荒漠和半荒漠地带的砂砾质坡地、石质丘陵顶部。
【用途】观赏。

Mongolian leathershell thistle
Aster family
Perennial herb, mucilaginous; stems basally dense-lanate, flowering stems 2～4 cm long; leaves basal, rosulate, leathery, pinnately parted to lobed, lobes crinkled, shallowly toothed and stiffly spinose, arachnoid-hairy; head solitary and terminal, staminate head smaller; outer involucral bracts leathery; tubular florets white. Flowering May to June.
Distribution: C Inner Mongolia; Mongolia
Habitat: Gravelly slopes and rocky hilltops in desert to semi-desert zones
Use: Ornamental

东方泽泻
Alisma orientale (Sam.) Juz.
泽泻科
Alismataceae

【特征】多年生水生或沼生草本；叶基生，叶片卵形或椭圆形，具长柄；花茎高30～100cm，花序分枝轮生，每轮3至多数；花直径3～5mm，花被片6，外轮绿色，内轮白色或粉色；瘦果多数，聚集成头状。花期6～8月。
【分布】东北、华北、西北、华东；蒙古、朝鲜、俄罗斯（远东）。
【生境】水边、沼泽。
【用途】药用。

Oriental water plantain
Water plantain family
Perennial aquatic or paludose herb; leaves basal, leaf blades ovate or elliptic, long-petiolate; scapes 30～100 cm tall, inflorescence with branches in whorls of 3 to several; flowers 3～5 mm across, tepals 6, the outer green, the inner white or pink; achenes numerous, congregated into a head. Flowering June to August.
Distribution: NE, N, NW and E China; Mongolia, Korea, Russia (Far East)
Habitat: Watersides and marshes
Use: Medicine

上图拍摄人：赵利清 Upper photo by Zhao Liqing

阿尔泰葱
Allium altaicum Pall.
百合科
Liliaceae

【特征】多年生草本；鳞茎卵状圆柱形；叶圆筒状，中空，粗0.5～2cm；花葶粗壮，高40cm以上，圆筒状，中空；总苞2裂，膜质；伞形花序球状，花多而密集；花被片6，黄白色；花丝长于花被片1.5～2倍。花期8月。
【分布】黑龙江、内蒙古、新疆；蒙古、西伯利亚、哈萨克斯坦。
【生境】石质山坡和草地。
【用途】饲用；食用；药用。

Altai onion
Lily family
Perennial herb; bulbs ovoid-cylindric; leaves terete, hollow, 0.5～2 cm wide; scapes stout, over 40 cm tall, terete, hollow; spathe 2-divided, membranous; umbel globose, with numerous and dense flowers; tepals 6, yellowish-white; filaments 1.5～2 times longer than the tepals. Flowering August.
Distribution: Heilongjiang, Inner Mongolia and Xinjiang; Mongolia, Siberia, Kazakhstan
Habitat: Rocky slopes and grasslands
Use: Forage; edible; medicine

茗葱
Allium victorialis L.
百合科
Liliaceae

【特征】多年生草本；鳞茎多单生，近圆柱状；叶2～3，倒披针状椭圆形至椭圆形；花葶高25～80cm；总苞2裂；伞形花序球状，花多而密集；花被片6，白色；花丝比花被片长达1倍。花期6～7月。

【分布】东北、华北、华中、陕西、甘肃、四川；北温带地区。

【生境】海拔1 000～2 500m的山地林下、林缘、草甸、沟边。

【用途】饲用；食用；药用。

Longroot chive (Alpine leek, Victory onion)
Lily family
Perennial herb; bulbs usually solitary, subcylindric; leaves 2～3, oblanceolate-elliptic to elliptic; scapes 25～80 cm tall; spathe 2-divided; umbel globose, with numerous and dense flowers; tepals 6, white; filaments up to once longer than the tepals. Flowering June to July.
Distribution: NE, N and C China, Shaanxi, Gansu and Sichuan; the northern temperate zone
Habitat: Montane forests, forest margins, meadows and ditch sides at 1000～2 500 m
Use: Forage; edible; medicine

拍摄人：赵利清 Photos by Zhao Liqing

右图拍摄人：赵利清 Right photo by Zhao Liqing

黄精
Polygonatum sibiricum Redouté
百合科
Liliaceae

【特征】多年生草本，高50～90cm；茎直立或近攀援；叶4～6轮生，先端拳卷或钩状；花2～4枚腋生；花梗下垂，基部有膜质苞片；花被筒状钟形，白色或淡黄色，中部稍缢缩；浆果熟时黑色。花期5～6月。
【分布】东北、华北、华东、陕西、宁夏、甘肃；蒙古、朝鲜、西伯利亚。
【生境】林下、灌丛、山地草甸。
【用途】饲用；药用。

Siberian solomon's seal
Lily family
Perennial herb 50～90 cm tall; stems erect or subscandent; leaves in whorls of 4～6, apex circinate or hamate; flowers 2～4, axillary; pedicels drooping, base with membranous bracts; perianth tubular-campanulate, white or pale yellow, slightly constricted in middle; berries black at maturity. Flowering May to June.
Distribution: NE, N and E China, Shaanxi, Ningxia and Gansu; Mongolia, Korea, Siberia
Habitat: Forests, thickets and montane meadows
Use: Forage; medicine

拍摄人：赵 凡 Photos by Zhao Fan

牛扁
Aconitum barbatum var. *puberulum* Ledeb.
毛茛科
Ranunculaceae

【特征】多年生草本，高达1m；茎直立，被贴伏的短柔毛；叶掌状3全裂，裂片羽状分裂，末回小裂片披针形或狭卵形，两面被柔毛；总状花序花密集；萼片5，黄色，盔瓣圆筒形。花期7~8月。
【分布】华北，新疆东部；西伯利亚。
【生境】山地、沟谷，疏林、灌丛。
【用途】有毒植物。根药用。

Ox monkshood
Buttercup family
Perennial herb up to 1 m tall; stems erect, appressed-pubescent; leaves palmately 3-divided, segments pinnately parted, ultimate segment lanceolate or narrowly ovate, pubescent; racemes with dense flowers; sepals 5, yellow, galea cylindric. Flowering July to August.
Distribution: N China, E Xinjiang; Siberia
Habitat: Mountains and ravines, open woodlands and thickets
Use: Poisonous. Roots for medicine

灰叶铁线莲
Clematis canescens (Turcz.) W. T. Wang et M. C. Chang
毛茛科
Ranunculaceae

【特征】灌木，高达1m；茎直立；单叶对生或簇生，革质，灰绿色，全缘，稀基部具齿或小裂片，两面被细柔毛；花单生或聚伞花序具3花；萼片4，斜上展，黄色，边缘外面密被绒毛；无花瓣；瘦果多数，密被长柔毛，宿存花柱羽毛状。花期7～8月。
【分布】内蒙古西部、宁夏、甘肃北部。
【生境】荒漠和荒漠草原地带的山地、石质残丘、沙地和沙丘间洼地。
【用途】饲用；观赏。

Hoary clematis
Buttercup family
Shrub to 1 m tall; stems erect; simple leaves opposite or clustered, leathery, grey-green, entire, rarely toothed or lobed at base, puberulous; flowers solitary or 3 in a cyme; sepals 4, ascending, yellow, margins densely tomentose outer side; petals without; achenes numerous, densely villous, persistent styles plumose. Flowering July to August.
Distribution: W Inner Mongolia, Ningxia and N Gansu
Habitat: Mountains, rocky monadnocks, sands, and swales among dunes in desert to desert-steppe zones
Use: Forage; ornamental

拍摄人：赵利清 Photos by Zhao Liqing

鄂尔多斯小檗
Berberis carolii Schneid.
小檗科
Berberidaceae

【特征】灌木，高1~2m；幼枝紫褐色；叶刺单一或3分叉，长10~25mm；叶簇生于刺腋，倒卵状披针形至椭圆形，全缘或有刺齿，下面被白粉；总状花序稍下垂，花9~15枚；萼片6，黄色；花瓣6，黄色；浆果矩圆状，鲜红色。花期5~6月。
【分布】内蒙古中部和西部。
【生境】草原带山地。
【用途】药用；水土保持。

Ordos barberry
Barberry family
Shrub 1~2 m tall; young branchlets purple-brown; thorn single or 3-forked, 10~25 mm long; leaves clustered in thorn axils, obovate-lanceolate to elliptic, entire or spinose-toothed, farinose beneath; racemes slightly nodding, with 9~15 flowers; sepals 6, yellow; petals 6, yellow; berries oblong, bright-red. Flowering May to June.
Distribution: C and W Inner Mongolia
Habitat: Mountains in steppe zone
Use: Medicine; soil conservation

细叶小檗
Berberis poiretii Schneid.
小檗科
Berberidaceae

【特征】灌木，高1～2m；幼枝紫褐色；叶刺单一或3～5分叉，长4～9mm；叶簇生于刺腋，倒披针形至披针状匙形，全缘或中上部边缘有刺齿；总状花序下垂，花8～15枚；萼片6，黄色；花瓣6，黄色；浆果矩圆状，鲜红色。花期5～6月。

【分布】东北、华北；蒙古、朝鲜、俄罗斯（远东）。

【生境】砾石质山坡、坡麓、残丘、沙地。

【用途】药用；水土保持。

Slenderleaf barberry
Barberry family
Shrub 1～2 m tall; young branchlets purple-brown; thorn single or 3- to 5-forked, 4～9 mm long; leaves clustered in thorn axils, oblanceolate to lanceolate-spatulate, entire or spinose-toothed in upper half; racemes nodding, with 8～15 flowers; sepals 6, yellow; petals 6, yellow; berries oblong, bright-red. Flowering May to June.
Distribution: NE and N China; Mongolia, Korea, Russia (Far East)
Habitat: Gravelly slopes, hillsides, monadnocks and sands
Use: Medicine; soil conservation

左图拍摄人：赵利清 Left photo by Zhao Liqing

拍摄人：赵利清 Photo by Zhao Liqing

白屈菜
Chelidonium majus L.
罂粟科
Papaveraceae

【特征】多年生草本，高30～80cm，含橙色乳汁；茎直立，具分枝，密被毛；叶羽状全裂，裂片羽状浅裂或具圆齿；伞形花序顶生和腋生；萼片2，疏被毛；花瓣4，2轮，黄色；雄蕊多数；柱头2浅裂。花期5～7月。
【分布】东北、华北、西北、华东、华中、西南；蒙古、朝鲜、日本、俄罗斯、欧洲。
【生境】山地林下、林缘、草甸、溪边。
【用途】全株有毒。药用。

Greater celandine
Poppy family
Perennial herb 30～80 cm tall, with orange milky juice; stems erect, branched, densely hairy; leaves pinnatisect, segments pinnatilobate or crenate; umbels terminal and axillary; sepals 2, sparsely hairy; petals 4, in 2 series, yellow; stamens numerous; stigmas 2-lobed. Flowering May to July.
Distribution: NE, N, NW, E, C and SW China; Mongolia, Korea, Japan, Russia, Europe
Habitat: Montane forests, forest margins, meadows, and stream banks
Use: Plants poisonous throughout. Medicine

杂交景天
Sedum hybridum L.
景天科
Crassulaceae

【特征】多年生肉质草本；茎斜升，具匍匐茎，不育枝短，花茎高达30cm；叶互生，匙状椭圆形至倒卵形，边缘有钝锯齿；花序聚伞状，顶生；花瓣5，黄色，披针形；蓇葖果成熟后星芒状开展。花期6～7月。
【分布】新疆北部；蒙古、俄罗斯、中亚。
【生境】海拔730～2 700m的山沟水边、山谷阴处、碎石质草地、山坡石缝。
【用途】庭院绿化。

Hybrid stonecrop
Stonecrop family
Perennial fleshy herb; stems ascending, stoloniferous, sterile shoots short, flowering stems to 30 cm tall; leaves alternate, spatulate-elliptic to obovate, obtusely serrate; inflorescence cymose, terminal; petals 5, yellow, lanceolate; follicles stellately divergent at maturity. Flowering June to July.
Distribution: N Xinjiang; Mongolia, Russia, Central Asia
Habitat: Gully watersides, shady places in valleys, broken-rocky grasslands and sloping rock crevices at 730～2 700 m
Use: Courtyard planting

拍摄人：卢欣石　Photos by Lu Xinshi

拍摄人：赵利清 Photos by Zhao Liqing

委陵菜
Potentilla chinensis Ser.
蔷薇科
Rosaceae

【特征】多年生草本，高20～50cm；茎直立或斜升，被柔毛；单数羽状复叶，小叶片羽状半裂或深裂，边缘反卷，上面被短柔毛，下面被白色毡毛；伞房状聚伞花序，花多数；花梗和花萼被柔毛；花瓣5，黄色。花期7～8月。
【分布】东北至西北及西南大部分省区；蒙古、朝鲜、日本、俄罗斯。
【生境】草原、灌丛、草甸。
【用途】饲用；药用。

Chinese cinquefoil
Rose family
Perennial herb 20～50 cm tall; stems erect or ascending, pilose; leaves odd-pinnate, leaflet blades pinnately cleft or parted, margins revolute, pubescent above, white-manicate below; corymbose cymes with many flowers; pedicel and calyx pilose; petals 5, yellow. Flowering July to August.
Distribution: Most regions from NE to NW and SW China; Mongolia, Korea, Japan, Russia
Habitat: Grasslands, scrublands and meadows
Use: Forage; medicine

多茎委陵菜
Potentilla multicaulis **Bunge**
蔷薇科
Rosaceae

【特征】多年生草本，高10～25cm；茎丛生，斜升或斜倚，密被柔毛；单数羽状复叶，基生叶多数，小叶片羽状深裂，上面被柔毛，下面密被白色毡毛；伞房状聚伞花序，花少数；花瓣5，黄色。花期6～7月。
【分布】东北至西北及西南；蒙古、俄罗斯。
【生境】草地、田边。
【用途】饲用；药用。

Many-stemmed cinquefoil
Rose family
Perennial herb 10～25 cm tall; stems tufted, ascending or decumbent, densely pilose; leaves odd-pinnate, basal leaves abundant, leaflet blades pinnatipartite, pubescent above, densely white-manicate below; corymbose cymes with few flowers; petals 5, yellow. Flowering June to July.
Distribution: NE to NW and SW China; Mongolia, Russia
Habitat: Grasslands and farmland sides
Use: Forage; medicine

掌叶多裂委陵菜
Potentilla multifida var. *ornithopoda* (Tausch) Th. Wolf
蔷薇科
Rosaceae

【特征】多年生草本，高20～40cm；茎斜升、斜倚或近直立，被柔毛；单数羽状复叶，小叶5，排列紧密，似掌状复叶，小叶片羽状深裂，裂片条形或条状披针形，边缘反卷，下面被白色毡毛；伞房状聚伞花序顶生；花瓣5，黄色。花期7～8月。

【分布】华北，黑龙江、陕西、甘肃、新疆、西藏；蒙古、俄罗斯。

【生境】草地。

【用途】饲用。

Palmate staghorn cinquefoil
Rose family
Perennial herb 20～40 cm tall; stems ascending, decumbent or suberect, pubescent; leaves odd-pinnate, leaflets 5, closely arranged, similarly palmate, leaflet blades pinnatipartite, segments linear or linear-lanceolate, margins revolute, white-manicate below; corymbose cymes terminal; petals 5, yellow. Flowering July to August.
Distribution: N China, Heilongjiang, Shaanxi, Gansu, Xinjiang and Tibet; Mongolia, Russia
Habitat: Grasslands
Use: Forage

拍摄人：赵利清 Photos by Zhao Liqing

西山委陵菜
Potentilla sischanensis Bunge ex Lehm.
蔷薇科
Rosaceae

【特征】多年生草本，高7～20cm；茎丛生，直立或斜升，被毡毛；单数羽状复叶，多基生，小叶7～13，近革质，羽状深裂，顶生3小叶较大，裂片全缘，边缘反卷，下面密被白色毡毛；聚伞花序，花少数；花萼被毡毛；花瓣5，黄色。花期5～8月。

【分布】华北、陕西、宁夏、甘肃、青海、四川；蒙古。

【生境】山地阳坡、石质丘陵。

【用途】饲用。

Xishan cinquefoil
Rose family
Perennial herb 7～20 cm tall; stems tufted, erect or ascending, manicate; leaves odd-pinnate, mostly basal, leaflets 7～13, subleathery, pinnatipartite, the terminal 3 larger, segments entire, margins revolute, densely white-manicate below; cymes with few flowers; calyx manicate; petals 5, yellow. Flowering May to August.
Distribution: N China, Shaanxi, Ningxia, Gansu, Qinghai and Sichuan; Mongolia
Habitat: Mountain sunny slopes and rocky hills
Use: Forage

黄耆
Astragalus membranaceus **Bunge**
豆科
Fabaceae (Leguminosae)

【特征】多年生草本，高50～100cm；茎直立或斜升，多分枝；单数羽状复叶，小叶13～27，下面被柔毛；总状花序腋生，具花10～20枚；蝶形花冠黄色或淡黄色；荚果膨胀，薄膜质，被柔毛。花期6～8月。

【分布】东北、华北、西北；蒙古、西伯利亚。

【生境】草甸、灌丛、疏林或林缘。

【用途】饲用；绿肥；根药用。

Membranous milkvetch
Pea family
Perennial herb 50～100 cm tall; stems erect or ascending, much branched; leaves odd-pinnate, leaflets 13～27, pubescent beneath; racemes axillary with 10～20 flowers; papilionaceous corolla yellow or pale yellow; pods inflated, thinly membranous, pubescent. Flowering June to August.
Distribution: NE, N and NW China; Mongolia, Siberia
Habitat: Meadows, thickets, open woodlands and forest margins
Use: Forage; green manure; roots for medicine

上图拍摄人：赵利清 Upper photo by Zhao Liqing

长毛荚黄耆
Astragalus monophyllus Maxim.
豆科
Fabaceae (Leguminosae)

【特征】多年生矮小草本，高3～5cm；地上茎无或极短缩；三出复叶基生，小叶1或3，宽卵形或近圆形，被白色伏贴毛；总状花序具1～2花；蝶形花冠淡黄色或近白色；荚果矩圆状至矩圆状卵形，稍膨胀，具喙，密被白绵毛。花期4～5月。
【分布】内蒙古中西部、山西、甘肃、新疆；蒙古。
【生境】荒漠和荒漠草原地带的砾石质山坡、戈壁。
【用途】饲用。

Long woolypod milkvetch
Pea family
Perennial dwarf herb 3～5 cm tall; aucalescent or stems strongly shortened; leaves ternate, basal, leaflets 1 or 3, blades broadly ovate or suborbicular, white appressed-hairy; racemes with 1～2 flowers; papilionaceous corolla pale yellow or nearly white; pods oblong to oblong-ovoid, slightly inflated, beaked, densely white-lanate. Flowering April to May.
Distribution: CW Inner Mongolia, Shanxi, Gansu and Xinjiang; Mongolia
Habitat: Gravelly slopes and gobi in desert and desert-steppe zones
Use: Forage

湿地黄耆
Astragalus uliginosus L.
豆科
Fabaceae (Leguminosae)

【特征】多年生草本，高30～60cm；茎单一或数个，直立，被伏贴毛；单数羽状复叶，小叶13～27，小叶片椭圆形至矩圆形，上面无毛，下面被伏贴毛；总状花序，花密集，下斜；蝶形花冠淡黄色；荚果矩圆状，膨胀，具喙，果皮革质，无毛。花期6～8月。

【分布】东北，内蒙古；蒙古、朝鲜、俄罗斯（西伯利亚、远东）。

【生境】草甸、沼泽草甸、河边。

【用途】饲用。

Marsh milkvetch
Pea family
Perennial herb 30～60 cm tall; stems single or several, erect, appressed-hairy; leaves odd-pinnate, leaflets 13～27, blades elliptic to oblong, glabrous above, appressed-hairy below; racemes with dense and descending flowers; papilionaceous corolla pale yellow; pods oblong, inflated, beaked, pericarp leathery, glabrous. Flowering June to August.
Distribution: NE China, Inner Mongolia; Mongolia, Korea, Russia (Siberia and Far East)
Habitat: Meadows, marsh-meadows and riversides
Use: Forage

拍摄人：赵利清 Photo by Zhao Liqing

宽叶多序岩黄耆
Hedysarum polybotrys var. *alaschanicum*
(B. Fedtsch.) H.C. Fu et Z.Y. Chu
豆科
Fabaceae (Leguminosae)

【特征】多年生草本，高达1m；茎直立，稍分枝；单数羽状复叶，小叶7～25，小叶片卵形至椭圆形，上面无毛，下面中脉被长柔毛；总状花序腋生，花20～25枚；蝶形花冠淡黄色；荚果3～5节，扁平，边缘具狭翅。花期6～7月。
【分布】河北、内蒙古、宁夏。
【生境】草原和荒漠区的山坡、沟谷或林缘。
【用途】饲用；药用。

Broadleaf grapecluster sweetvetch
Pea family
Perennial herb to 1 m tall; stems erect, slightly branched; leaves odd-pinnate, leaflets 7～25, blades ovate to elliptic, glabrous above, villous on midvein beneath; racemes axillary with 20～25 flowers; papilionaceous corolla pale yellow; pods with 3～5 loments, flat, margins narrowly winged. Flowering June to July.
Distribution: Hebei, Inner Mongolia and Ningxia
Habitat: Slopes, ravines or forest edges in steppe and desert areas
Use: Forage; medicine

牧地山黧豆
Lathyrus pratensis L.
豆科
Fabaceae (Leguminosae)

【特征】多年生草本，高30～120cm；茎上升、平卧或攀援；叶具1对小叶，叶轴末端具卷须，卷须单一或分枝，小叶片椭圆形至条状披针形，脉平行；总状花序腋生，花5～12枚；蝶形花冠黄色；荚果条状。花期6～8月。
【分布】黑龙江、陕西、甘肃、青海、新疆、四川、云南、贵州；欧亚大陆广布。
【生境】海拔1 000～3 000m的疏林、灌丛、草甸。
【用途】饲用；绿肥；蜜源。

Meadow pea (Meadow vetch)
Pea Family
Perennial herb 30～120 cm tall; stems ascending, procumbent or climbing; leaves with leaflets in 1 pairs, rachis terminating in a solitary or branched tendril, blades elliptic to linear-lanceolate, parallel-veined; racemes axillary with 5～12 flowers; papilionaceous corolla yellow; pods linear. Flowering June to August.
Distribution: Heilongjiang, Shaanxi, Gansu, Qinghai, Xinjiang, Sichuan, Yunnan and Guizhou; widespread in Eurasia
Habitat: Open woodlands, thickets and meadows at 1 000～3 000 m
Use: Forage; green manure; honey source

拍摄人：卢欣石 Photos by Lu Xinshi

蒺藜
Tribulus terrestris L.
蒺藜科
Zygophyllaceae

【特征】一年生草本；茎平铺地面，长达1m，由基部分枝；偶数羽状复叶，小叶片矩圆形；花单生于叶腋；花瓣5，黄色；果由5个分果瓣组成，具针刺、短硬毛和瘤状突起。花期5~8月。
【分布】全国；全球温带地区。
【生境】山坡、沙地、荒地、田间、路旁、干河床、居民点附近。
【用途】饲用；药用。

Puncturevine
Caltrop family
Annual herb; stems prostrate to 1 m long, branched from base; leaves even-pinnate, leaflet blades oblong; flower solitary in leaf axils; petals 5, yellow; fruit with 5 mericarps, spinose, hispidulous and tuberculate. Flowering May to August.
Distribution: Throughout China; temperate regions in the world
Habitat: Slopes, sands, wastelands, farmlands, roadsides, dry riverbeds and residential places
Use: Forage; medicine

拍摄人：赵利清 Photos by Zhao Liqing

针枝芸香
Haplophyllum tragacanthoides Diels
芸香科
Rutaceae

【特征】小半灌木，高2～8cm；茎基部丛生针刺状老枝，当年枝直立、斜升或近于平卧；单叶互生，灰绿色，具腺点；花单生于枝顶；花瓣5，黄色；蒴果3瓣裂。花期6～7月。
【分布】内蒙古。
【生境】荒漠至半荒漠地带的石质山坡。
【用途】饲用。

Needletwig rue
Citrus family
Shrublet 2～8 cm tall; stems with tufted, acerose and old branches at base, new branches erect, ascending or nearly prostrate; simple leaves alternate, grey-green, glandular; solitary flower terminal; petals 5, yellow; capsules 3-valved. Flowering June to July.
Distribution: Inner Mongolia
Habitat: Rocky slopes in desert to semi-desert zones
Use: Forage

水金凤
Impatiens noli-tangere L.
凤仙花科
Balsaminaceae

【特征】一年生草本，高30~60cm；茎直立，上部分枝，肉质；叶互生，卵形至卵状披针形，边缘具疏钝齿；总状花序具花2~4枚；花两侧对称，2型；大花黄色，有时具紫红色斑点，下萼片瓣状，漏斗形，基部具长距；小花为闭锁花，淡黄色，无距。花期7~8月。

【分布】东北、华北、西北、华中；蒙古、朝鲜、日本、俄罗斯、欧洲。

【生境】林下、林缘湿地。

【用途】药用。

Touch-me-not impatiens
Touch-me-not family
Annual herb 30~60 cm tall; stems erect, branched above, fleshy; leaves alternate, ovate to ovate-lanceolate, sparsely blunt-toothed; racemes with flowers 2~4; flowers zygomorphic, dimorphic; large flowers yellow, sometimes with purple-red spots, lower sepal petal-like, funnelform, with long-spur at base; small flowers cleistogamous, pale yellow, spur without. Flowering July to August.

Distribution: NE, N, NW and C China; Mongolia, Korea, Japan, Russia, Europe
Habitat: Forests and wet sites along forest edges
Use: Medicine

拍摄人：赵利清 Photos by Zhao Liqing

蒙椴
Tilia mongolica Maxim.
椴树科
Tiliaceae

【特征】乔木，高达10m；树皮灰褐色，幼枝淡红褐色，光滑无毛；单叶互生，叶片宽卵形或近圆形，中上部常3裂，边缘具粗齿；聚伞花序下垂；花瓣5，黄色；退化雄蕊花瓣状，雄蕊多数，成5束；果实坚果状，被绒毛。花期7~8月。
【分布】华北；蒙古。
【生境】草原带山地。
【用途】树干为家具用材；花药用；种子提炼工业用油。

Mongolian basswood
Linden family
Tree to 10 m tall; barks grey-brown, twigs pale reddish-brown, glabrous; simple leaves alternate, blades broadly ovate or suborbicular, usually 3-lobed in upper half, margins coarsely serrate; cymes drooping; petals 5, yellow; staminodia petal-like, numerous stamens in 5 clusters; fruits nutlike, tomentose. Flowering July to August.
Distribution: N China; Mongolia
Habitat: Mountains in steppe zone
Use: Trunks for furniture materials; flowers for medicine; seeds for extracting industrial oil

野西瓜苗
Hibiscus trionum L.
锦葵科
Malvaceae

【特征】一年生草本，高10~60cm；茎直立或下部分枝铺散，被星状毛；叶掌状3~7半裂至全裂，裂片羽裂或具缺刻，下面被星状毛；花单生于叶腋；花萼膜质，果期膨大，具紫色脉纹，沿脉密生叉状硬毛；花瓣5，淡黄色或近白色，基部深紫色。花期6~9月。
【分布】全国各地；世界各地。
【生境】田野、路边、村庄附近。
【用途】药用。

Flower-of-an-hour (Venice mallow)
Mallow family
Annual herb 10~60 cm tall; stems erect or lower branches diffuse, stelipilous; leaves palmately 3- to 7-cleft to divided, segments pinnatilobate or incised, stelipilous beneath; flowers solitary in axils; calyx membranous, inflated in fruit, with purple veins, densely forked-hispid on veins; petals 5, pale yellow or nearly white, with the base dark purple. Flowering June to September.
Distribution: Throughout China; cosmopolitan
Habitat: Fields, roadsides and places around villages
Use: Medicine

拍摄人：赵 凡 Photos by Zhao Fan

黄海棠 (红旱莲)
Hypericum ascyron L.
金丝桃科 (藤黄科)
Hypericaceae (Clusiaceae)

【特征】多年生草本，高50～130cm；茎直立，单一或丛生，具4棱；单叶对生，叶片披针形至矩圆状卵形；聚伞花序顶生，具花1至数枚；花冠金黄色，花瓣5，镰状弯曲；雄蕊极多数；花柱在中部以上5裂。花期6～8月。
【分布】东北，黄河和长江流域；朝鲜、日本、西伯利亚。
【生境】林缘、灌丛、草甸。
【用途】观赏；药用；提制栲胶。

Great St. Johnswort
St. John's-wort family
Perennial herb 50～130 cm tall; stems erect, single or tufted, quadrangular; simple leaves opposite, blades lanceolate to oblong-ovate; cymes terminal, with 1 to several flowers; corolla golden-yellow, petals 5, falcate-curved; stamens very numerous; styles 5-parted in upper half. Flowering June to August.
Distribution: NE China, regions along Yellow River and Changjiang River; Korea, Japan, Siberia
Habitat: Forest margins, scrublands and meadows
Use: Ornamental; medicine; extracting tannin

乌腺金丝桃 (赶山鞭)
Hypericum attenuatum Fisch. ex Choisy
金丝桃科 (藤黄科)
Hypericaceae (Clusiaceae)

【特征】多年生草本，高30～60cm，全株散生黑色腺点；茎数个丛生，直立，具2纵线棱；单叶对生，叶片长卵形、倒卵形或椭圆形；聚伞状圆锥花序顶生，花数枚；花冠黄色，花瓣5，矩圆形或倒卵形；雄蕊多数；花柱3，自基部分离。花期7～8月。
【分布】东北、华北、华东、中南；蒙古、朝鲜、日本、俄罗斯（西伯利亚、远东）。
【生境】林下、林缘、灌丛、草甸、草原。
【用途】观赏；药用。

Lesser St. Johnswort
St. John's-wort family
Perennial herb 30～60 cm tall, with scattered black-glands throughout; stems several, tufted, erect, with 2 string ribs; simple leaves opposite, blades long-ovate, obovate or elliptic; thyrse terminal, with several flowers; corolla yellow, petals 5, oblong to obovate; stamens numerous; styles 3, divided from base. Flowering July to August.
Distribution: NE, N, E and CS China; Mongolia, Korea, Japan, Russia (Siberia and Far East)
Habitat: Forests, forest margins, scrublands, meadows and steppe
Use: Ornamental; medicine

拍摄人：赵利清 Photo by Zhao Liqing

拍摄人：赵利清 Photos by Zhao Liqing

双花堇菜
Viola biflora L.
堇菜科
Violaceae

【特征】多年生草本，高10～20cm；茎纤细，直立或上升，不分枝；叶片肾形，少近圆形，边缘具齿；花两侧对称，1～2枚腋生；花瓣5，黄色具紫色脉纹，最下瓣片基部具距，距长2～2.5mm。花果期5～9月。

【分布】东北、华北、西北、中南、西南；朝鲜、日本、俄罗斯、印度、巴基斯坦、欧洲、北美洲。

【生境】海拔2 500～4 000m的山地、亚高山和高山带疏林、林缘、灌丛、草甸、岩石缝。

【用途】饲用；药用。

Arctic yellow violet
Violet family
Perennial herb 10～20 cm tall; stems finely slender, erect or ascending, unbranched; leaf blades reniform, rarely suborbicular, serrate; flowers zygomorphic, 1 or 2 axillary; petals 5, yellow with purple streaks, the lower one with a basal spur 2～2.5 mm long. Flowering and fruiting May to September.

Distribution: NE, N, NW, CS and SW China; Korea, Japan, Russia, India, Pakistan, Europe, North America

Habitat: Open woodlands, forest edges, thickets, meadows and rock crevices in montane to subalpine and alpine zones at 2 500～4 000 m

Use: Forage; medicine

红柴胡
Bupleurum scorzonerifolium **Willd.**
伞形科
Apiaceae (Umbelliferae)

【特征】多年生草本，高10～60cm；主根红褐色；茎直立，常单一，稍呈"之"字形弯曲，基部具叶鞘残留纤维；叶条形或披针状条形，全缘；复伞形花序，伞幅4～18；小总苞片通常5，披针形；花瓣5，黄色。花期7～8月。
【分布】东北、华北、西北、华东；蒙古、朝鲜、日本、俄罗斯（西伯利亚、远东）。
【生境】草原、草甸、山坡灌丛、固定沙丘。
【用途】饲用；根药用。

Red thorow wax
Parsley family
Perennial herb 10～60 cm tall; taproot reddish-brown; stems erect, usually single, slightly flexuose, base clothed with fibrous remnant sheaths; leaves linear or lanceolate-linear, entire; compound umbels with 4～18 rays; bracteoles usually 5, lanceolate; petals 5, yellow. Flowering July to August.
Distribution: NE, N, NW and E China; Mongolia, Korea, Japan, Russia (Siberia and Far East)
Habitat: Steppe, meadows, sloping thickets and fixed dunes
Use: Forage; roots for medicine

兴安柴胡
***Bupleurum sibiricum* Vest ex Roem. et Schult.**
伞形科
Apiaceae (Umbelliferae)

【特征】多年生草本，高30～70cm；主根黑褐色；茎丛生，少单生，上部有分枝，基部具叶鞘残留纤维；叶条状倒披针形至狭披针形，全缘；复伞形花序，伞幅5～14；小总苞片5～12，花瓣状，椭圆状披针形至狭倒卵形，黄绿色，显著超出小伞形花序；花瓣5，黄色。花期7～8月。

【分布】黑龙江、辽宁、内蒙古；蒙古、西伯利亚。

【生境】山地草原、林缘草甸。

【用途】饲用；根药用。

Siberian thorow wax
Parsley family
Perennial herb 30～70 cm tall; taproot black-brown; stems tufted, rarely single, branched above, base clothed with fibrous remnant sheaths; leaves linear-oblanceolate to narrowly lanceolate, entire; compound umbels with 5～14 rays; bracteoles 5～12, petal-like, elliptic-lanceolate to narrowly obovate, yellowish-green, distinctly exceeding umbellet; petals 5, yellow. Flowering July to August.
Distribution: Heilongjiang, Liaoning and Inner Mongolia; Mongolia, Siberia
Habitat: Montane steppe, and meadows along forest edges
Use: Forage; roots for medicine

黑柴胡
Bupleurum smithii H. Wolff
伞形科
Apiaceae (Umbelliferae)

【特征】多年生草本，高25～60cm；主根红褐色；茎丛生，直立或斜升，基部无叶鞘残留纤维；基部叶矩圆状倒卵形，中部叶条形或倒披针形，上部叶卵形，全缘；复伞形花序，伞幅4～11；小总苞片6～9，花瓣状，卵形，黄绿色，长超过小伞形花序0.5～1倍；花瓣5，黄色。花期7～8月。

【分布】华北、陕西、甘肃、青海、河南。

【生境】山坡、山谷、山顶、草甸、灌丛。

【用途】饲用；根药用。

Black thorow wax
Parsley family
Perennial herb 25～60 cm tall; taproot red-brown; stems tufted, erect or ascending, base without fibrous remnant sheaths; basal leaves oblong-obovate, middle leaves linear or oblanceolate, upper leaves ovate, entire; compound umbels with 4～11 rays; bracteoles 6～9, petal-like, ovate, yellowish-green, 0.5～1 times longer than the umbellet; petals 5, yellow. Flowering July to August.
Distribution: N China, Shaanxi, Gansu, Qinghai and Henan
Habitat: Slopes, valleys, hilltops, meadows and thickets
Use: Forage; roots for medicine

拍摄人：赵 凡 Photos by Zhao Fan

沙茴香 (硬阿魏)
Ferula bungeana Kitag.
伞形科
Apiaceae (Umbelliferae)

【特征】多年生草本，高30～50cm；茎直立，具分枝；基生叶莲座状，3～4回羽状全裂，茎生叶向上渐小而简化，叶片质厚，通常灰蓝色，两面无毛；复伞形花序，伞幅5～15；花瓣5，黄色；果棱黄色。花期6～7月。
【分布】东北、华北、西北；蒙古。
【生境】草原和荒漠草原带沙地。
【用途】药用；饲用。

Coarse giant fennel
Parsley family
Perennial herb 30～50 cm tall; stems erect, branched; basal leaves rosulate, pinnatisect 3 to 4 times, cauline leaves gradually reduced upwards, blades thick, usually grey-blue, glabrous; compound umbels with 5～15 rays; petals 5, yellow; fruit ribs yellow. Flowering June to July.
Distribution: NE, N and NW China; Mongolia
Habitat: Sands in steppe and desert-steppe zones
Use: Medicine; forage

展毛黄芩
Scutellaria orthotricha C. Y. Wu et H. W. Li
唇形科
Lamiaceae (Labiatae)

【特征】半灌木，高10～15cm；茎多数，钝四棱形，密被柔毛；叶片卵形，边缘具圆齿；总状花序顶生；花梗和花萼被柔毛和腺毛；花冠淡黄色，带紫色斑，二唇形，外面被柔毛和腺毛，上唇盔状，先端微缺，下唇中裂片大。花期6～8月。
【分布】新疆北部。
【生境】山地草原、林地阳坡。
【用途】饲用。

Wooly skullcap
Mint family
Subshrub 10～15 cm tall; stems numerous, obtusely quadrangular, densely pubescent; leaf blades ovate, crenate; racemes terminal; pedicel and calyx pubescent and glandular-hairy; corolla pale yellow with purple spots, bilabiate, pubescent and glandular-hairy outside, upper lip galeate with retuse apex, the median lobe of lower lip large. Flowering June to August.
Distribution: N Xinjiang
Habitat: Montane steppe and wooded sunny slopes
Use: Forage

拍摄人：卢欣石 Photos by Lu Xinshi

天仙子
Hyoscyamus niger L.
茄科
Solanaceae

【特征】一年生或二年生草本，高30～80cm，全株密被黏性腺毛和柔毛，有气味；茎单一或分枝；基生叶莲座状，茎生叶互生，边缘羽状分裂，或具疏齿；花序总状或穗状；花冠钟状，5浅裂，黄色，有紫色网纹；蒴果卵球状。花期6～8月。

【分布】东北、华北、西北、西南；欧亚大陆、北非、北美洲。

【生境】田野、荒地、村舍附近、路旁。

【用途】药用；种子油制作肥皂和油漆。

Black henbane
Nightshade family
Annual or biennial herb 30～80 cm tall, densely viscid-piloglandulose and villous throughout, scented; stems single or branched; basal leaves rosulate, cauline leaves alternate, margins pinnately lobed or spaced-dentate; inflorescence racemose or spicate; corolla campanulate, 5-lobed, yellow with purple net-veins; capsules ovoid-globose. Flowering June to August.

Distribution: NE, N, NW and SW China; Eurasia, N Africa, North America
Habitat: Fields, wastelands, village sides and roadsides
Use: Medicine; seed oil for making soap and paint

秀丽马先蒿
Pedicularis venusta Schangin ex Bunge
玄参科
Scrophulariaceae

【特征】多年生草本，高10~40cm；茎直立，通常单一，不分枝，被卷毛；茎生叶互生，叶片羽状全裂，裂片羽状深裂；穗状花序顶生；萼钟状，近革质；花冠淡黄色或黄色，二唇形，上唇盔状，先端具2齿，下唇短于上唇，3浅裂。花期6~7月。

【分布】黑龙江、内蒙古、河北、新疆；蒙古、俄罗斯（西伯利亚、远东）。

【生境】山地和河谷，草甸、疏林和灌丛。

【用途】观赏。

Beautiful lousewort
Figwort family
Perennial herb 10~40 cm tall; stems erect, usually single, unbranched, curly; cauline leaves alternate, blades pinnatisect, segments pinnatipartite; spikes terminal; calyx campanulate, subleathery; corolla pale yellow or yellow, bilabiate, upper lip galeate with 2-toothed apex, lower lip shorter than the upper lip, 3-lobed. Flowering June to July.

Distribution: Heilongjiang, Inner Mongolia, Hebei and Xinjiang; Mongolia, Russia (Siberia and Far East)

Habitat: Mountains and valleys, meadows, open woodlands and thickets

Use: Ornamental

拍摄人：赵 凡 Photos by Zhao Fan

拍摄人：易 津 Photo by Yi Jin

黄花列当
Orobanche pycnostachya **Hance**
列当科
Orobanchaceae

【特征】根寄生植物，二年生或多年生草本，高10～40cm，全株密被腺毛；茎直立，圆柱形，不分枝；叶卵状披针形至条状披针形，干后黄褐色；穗状花序顶生；花冠二唇形，黄色，上唇2浅裂，下唇3浅裂；花药被长柔毛。花期6～7月。
【分布】东北、华北、华东；蒙古、朝鲜、俄罗斯（西伯利亚、远东）。
【生境】沙丘、山坡、草地；寄生在蒿属植物根上。
【用途】药用。

Yellow broomrape
Broomrape family
Root parasitic, biennial or perennial herb 10～40 cm tall, densely glandular-pubescent throughout; stems erect, terete, unbranched; leaves ovate-lanceolate to linear-lanceolate, yellow-brown when dry; spikes terminal; corolla bilabiate, yellow, upper lip 2-lobed, lower lip 3-lobed; anthers villous. Flowering June to July.
Distribution: NE, N and E China; Mongolia, Korea, Russia (Siberia and Far East)
Habitat: Dunes, slopes, grasslands; parasitic to roots of sages
Use: Medicine

异叶败酱 (墓头回)
Patrinia heterophylla Bunge
败酱科
Valerianaceae

【特征】多年生草本，高15～80cm；茎直立；基生叶丛生，不分裂或羽状深裂至全裂，茎生叶对生，羽状全裂；伞房状聚伞花序顶生；总苞叶条形，不裂或1～2对条裂；花冠黄色，管钟状，裂片5；瘦果具翅状干膜质苞片。花期7～9月。

【分布】华北、华东、河南、陕西、宁夏、甘肃、青海。

【生境】山地草原及林缘。

【用途】药用；饲用。

Rough patrinia
Valerian family
Perennial herb 15～80 cm tall; stems erect; basal leaves tufted, undivided or pinnately parted to divided, cauline leaves opposite, pinnately divided; corymbose cymes terminal; bracteal leaves linear, undivided or linear-parted in 1 to 2 pairs; yellow corolla tubular-campanulate with 5 lobes; achenes with wing-like and scarious bract. Flowering July to September.

Distribution: N and E China, Henan, Shaanxi, Ningxia, Gansu and Qinghai
Habitat: Montane steppe and forest margins
Use: Medicine; forage

拍摄人：赵利清 Photos by Zhao Liqing

败酱 (黄花龙芽)
Patrinia scabiosifolia Fisch. ex Trevir.
败酱科
Valerianaceae

【特征】多年生草本，高30～80 (150)cm；茎直立，黄绿色至黄褐色；基生叶丛生，不分裂或羽状分裂，茎生叶对生，羽状深裂或全裂；聚伞圆锥花序在顶端组成大型伞房花序，花序分枝一侧被毛；花冠黄色或带白色，钟状，裂片5；瘦果无翅状苞片。花期7～9月。
【分布】几遍全国；蒙古、朝鲜、日本、俄罗斯（西伯利亚、远东）。
【生境】林下、林缘、灌丛、草甸。
【用途】药用；饲用；幼苗嫩叶可食。

Yellow patrinia
Valerian family
Perennial herb 30～80 (150) cm tall; stems erect, yellowish-green to yellowish-brown; basal leaves tufted, undivided or pinnately parted, cauline leaves opposite, pinnately parted or divided; thyrsus forming a terminal and large corymb, branches hairy one side; corolla yellow or whitish, campanulate with 5 lobes; achenes without wing-like bract. Flowering July to September.
Distribution: Almost throughout China; Mongolia, Korea, Japan, Russia (Siberia and Far East)
Habitat: Forests, forest margins, thickets and meadows
Use: Medicine; forage; seedlings and young leaves edible

猫儿菊
Achyrophorus ciliatus (Thunb.) Sch. Bip.
(*Hypochaeris ciliata* (Thunb.) Makino)
菊科
Asteraceae (Compositae)

【特征】多年生草本，高20～60cm；茎直立，不分枝；叶椭圆形至长卵形，边缘具尖齿，两面被硬毛；头状花序单生于茎顶；总苞宽钟状或半球形，被毛；舌状小花多数，橘黄色，长达3cm。花期6～7月。
【分布】东北、华北；蒙古、朝鲜、日本、俄罗斯（西伯利亚、远东）。
【生境】山地林缘、草甸。
【用途】观赏；药用。

Common cat's ear
Aster family
Perennial herb 20～60 cm tall; stems erect, unbranched; leaves elliptic to long-ovate, margins pointed-toothed, hispid both sides; head solitary and terminal; involucre broadly campanulate or hemispheric, hairy; ligulate florets numerous, orange-yellow, to 3 cm long. Flowering June to July.
Distribution: NE and N China; Mongolia, Korea, Japan, Russia (Siberia and Far East)
Habitat: Montane forest margins and meadows
Use: Ornamental; medicine

灌木亚菊
Ajania fruticulosa (Ledeb.) Poljak.
菊科
Asteraceae (Compositae)

【特征】小半灌木，高10～40cm；茎直立或斜升，密被毛，上部分枝；叶灰绿色，2回掌状或掌式羽状3～5全裂，上部叶有时不分裂，两面被毛及腺点；头状花序3～25，伞房状排列；管状小花黄色，外面有腺点。花期8～9月。
【分布】西北、内蒙古西部、西藏；中亚、蒙古。
【生境】荒漠和荒漠草原带的砂砾质和石质坡地。
【用途】饲用；药用。

Shrubby ajania
Aster family
Shrublet 10～40 cm tall; stems erect or ascending, densely hairy, branched above; leaves grey-green, bipalmately or palmate-pinnately 3- to 5-divided, upper leaves sometimes undivided, hairy and glandular; heads 3～25, corymbiform-arranged; tubular florets yellow with glands outside. Flowering August to September.
Distribution: NW China, W Inner Mongolia and Tibet; Central Asia, Mongolia
Habitat: Gravelly and rocky slopes in desert and desert-steppe zones
Use: Forage; medicine

铺散亚菊
Ajania khartensis (Dunn) Shih
菊科
Asteraceae (Compositae)

【特征】多年生草本，高10～30cm，全株密被白色绢毛；茎铺散状；叶沿枝密集排列，2回掌状或近掌状3～5全裂，两面密被灰白色短柔毛；头状花序少数，复伞房状排列；总苞片边缘褐色膜质，背部密被长柔毛；管状小花黄色。花期8～9月。
【分布】西南，内蒙古西部、宁夏、甘肃、青海；中亚、蒙古。
【生境】荒漠草原地带的砾石质山坡和坡麓。
【用途】饲用。

Spreading ajania
Aster family
Perennial herb 10～30 cm tall, densely white-sericeous throughout; stems diffuse; leaves densely arranged on branches, bipalmately or subpalmately 3- to 5-divided, densely grey-white pubescent; few heads compound corymbiform-arranged; involucral bracts with brown-membranous edges and densely villous back; tubular florets yellow. Flowering August to September.

Distribution: SW China, W Inner Mongolia, Ningxia, Gansu and Qinghai; Central Asia, Mongolia
Habitat: Gravelly slopes and foothills in desert-steppe zone
Use: Forage

山蒿(岩蒿)
Artemisia brachyloba Franch.
菊科
Asteraceae (Compositae)

【特征】半灌木状草本或小半灌木,高20~40cm;茎多数,自基部分枝,形成球状株丛;基生叶2~3回羽状全裂,茎生叶1~2回羽状全裂,小裂片条形或狭条形,边缘反卷,下面密被绒毛;头状花序直径2~3.5mm,略下倾;总苞片被绒毛。花期8~9月。
【分布】华北,陕西、宁夏、甘肃;蒙古。
【生境】草原和荒漠草原地带的石质残丘、山地阳坡、岩石缝。
【用途】饲用;药用。

Cliffrock sagebrush
Aster family
Subfrutical herb or shrublet 20~40 cm tall; stems numerous, branched from base, forming a globose stand; basal leaves pinnatisect twice to thrice, cauline leaves pinnatisect once to twice, lobules linear or narrowly linear, margins revolute, densely tomentose below; heads 2~3.5 mm across, slightly declinate; involucral bracts tomentose. Flowering August to September.
Distribution: N China, Shaanxi, Ningxia and Gansu; Mongolia
Habitat: Rocky monadnocks, sunny slopes and rock crevices in steppe and desert-steppe zones
Use: Forage; medicine

漠蒿 (沙蒿)
Artemisia desertorum Spreng.
菊科
Asteraceae (Compositae)

【特征】多年生草本，高10～70cm；茎直立，单生或少数，上部有分枝；茎下部叶2回状羽状全裂至深裂，中部叶1～2回羽状深裂，上部叶3～5深裂；头状花序直径2.5～3mm；总苞片被微毛，后脱落无毛。花期7～8月。

【分布】东北、华北、西北、西南；蒙古、朝鲜、日本、俄罗斯（西伯利亚、远东）、巴基斯坦、印度。

【生境】草原、高山草原、草甸、林缘。

【用途】饲用。

Desert sage
Aster family
Perennial herb 10～70 cm tall; stems erect, single or few, branched above; lower cauline leaves bipinnately divided to parted, middle leaves pinnately to bipinnately parted, upper leaves 3- to 5-parted; heads 2.5～3 mm across; involucral bracts puberulent to glabrate. Flowering July to August.
Distribution: NE, N, NW and SW China; Mongolia, Korea, Japan, Russia (Siberia and Far East), Pakistan, India
Habitat: Steppe, alpine-steppe, meadows and forest margins
Use: Forage

拍摄人：赵利清 Photos by Zhao Liqing

褐沙蒿
Artemisia intramongolica H. C. Fu
菊科
Asteraceae (Compositae)

【特征】半灌木，高30～60cm；茎直立或斜升，自基部分枝，新枝褐色或紫褐色；下部叶2～3回羽状全裂，小裂片狭条形或丝状条形，叶柄长3～7cm，上部叶分裂渐少，无柄；头状花序卵形或长卵形，直径1.5～2mm，通常直立；总苞片无毛。花期7～9月。
【分布】内蒙古（浑善达克沙地）。
【生境】草原带的沙地和沙丘。
【用途】饲用；固沙。

Inner Mongolian sagebrush
Aster family
Subshrub 30～60 cm tall; stems erect or ascending, branched from base, new branches brown or purple-brown; lower leaves pinnatisect twice to thrice, lobules narrowly linear or filiform-linear, petiole 3～7 cm long, upper leaves reduced and sessile; heads ovoid or long-ovoid, 1.5～2 mm across, usually erect; involucral bracts glabrous. Flowering July to September.
Distribution: Inner Mongolia (Hunshandake Sands)
Habitat: Sands and dunes in steppe zone
Use: Forage; fixing dunes

蒙古蒿
Artemisia mongolica (Fisch. ex Besser) Nakai
菊科
Asteraceae (Compositae)

【特征】多年生草本，高20～90cm；茎直立，少数或单生，常带紫褐色，多分枝；下部叶2回羽状全裂或深裂，上面绿色，下面密被灰白色蛛丝状毛，上部叶3～5全裂；头状花序椭圆形，直径1.5～2mm，直立或斜倾；总苞片密被蛛丝状毛。花期8～9月。
【分布】东北、华北、西北、华东、华中、东南；蒙古、朝鲜、日本、西伯利亚。
【生境】草原、草甸、沙丘、撂荒地、农田、路旁。
【用途】饲用；药用；提取芳香油；纤维材料；造纸。

Mongolian sage
Aster family
Perennial herb 20～90 cm tall; stems erect, few or single, usually purplish-brown, much branched; lower leaves bipinnately divided or parted, green above, densely grey-white arachnoid-pubescent below, upper leaves 3- to 5-divided; heads ellipsoid, 1.5～2 mm across, erect or ascending; involucral bracts densely arachnoid-pubescent. Flowering August to September.
Distribution: NE, N, NW, E, C and SE China; Mongolia, Korea, Japan, Siberia
Habitat: Steppe, meadows, dunes, abandoned lands, farmlands and roadsides
Use: Forage; medicine; extracting essential oil; fiber materials; papermaking

黑蒿 (沼泽蒿)
Artemisia palustris L.
菊科
Asteraceae (Compositae)

【特征】一年生草本，高10~40cm，全株光滑无毛；茎单生，直立，绿色，有时带紫褐色；叶1~2回羽状全裂，小裂片狭条形，上部叶渐小；头状花序近球形，直径2~3mm，无梗，每2~10枚密集成簇；总苞片无毛。花期8~9月。
【分布】东北，内蒙古、河北北部；蒙古、朝鲜、俄罗斯（西伯利亚、远东）。
【生境】森林至草原带的沙地、草甸、河岸。
【用途】饲用。

Black sage
Aster family
Annual herb 10~40 cm tall, glabrous throughout; stems single, erect, green or purplish-brown; leaves pinnatisect once to twice, lobules narrowly linear, upper leaves reduced; heads subglobose, 2~3 mm across, sessile, 2~10 in a cluster; involucral bracts glabrous. Flowering August to September.
Distribution: NE China, Inner Mongolia and N Hebei; Mongolia, Korea, Russia (Siberia and Far East)
Habitat: Sands, meadows and river banks in forest to steppe zones
Use: Forage

白莲蒿 (万年蒿、铁杆蒿)
Artemisia sacrorum Ledeb.
菊科
Asteraceae (Compositae)

【特征】半灌木状草本，高50~100cm；茎多数，褐色，多分枝；下部叶2~3回栉齿状羽状全裂，叶中轴两侧有栉齿，叶上面绿色，下面初时密被柔毛，后无毛，上部叶渐小；头状花序近球形，下垂；总苞片被柔毛，后脱落无毛。花期8~9月。

【分布】除高寒地区外，几遍全国；蒙古、朝鲜、日本、俄罗斯（西伯利亚、远东）、中亚、阿富汗、巴基斯坦、印度、尼泊尔。

【生境】山坡、路旁。

【用途】饲用；药用。

Russian wormwood
Aster family
Subfrutical herb 50~100 cm tall; stems numerous, brown, much branched; lower leaves pectinately pinnatisect twice to thrice, middle rachis with pectinate lobes both sides, green above, densely pubescent to glabrate below, upper leaves reduced; heads subglobose, nodding; involucral bracts pubescent to glabrate. Flowering August to September.

Distribution: Almost throughout China except the alpine cold areas; Mongolia, Korea, Japan, Russia (Siberia and Far East), Central Asia, Afghanistan, Pakistan, India and Nepal

Habitat: Mountains slopes and roadsides

Use: Forage; medicine

密毛白莲蒿
Artemisia sacrorum var. *messerschmidtiana* (Bess.) Y. R. Ling
菊科
Asteraceae (Compositae)

【特征】与白莲蒿(*Artemisia sacrorum*)的区别为：叶两面密被灰白色或灰黄色短柔毛。
【分布】东北、华北、西北，山东、江苏、河南；蒙古、朝鲜、日本、西伯利亚。
【生境】山坡、路旁。
【用途】饲用；药用。

Wooly Russian wormwood
Aster family
Difference to *Artemisia sacrorum*: Leaves densely grey-white or grey-yellow pubescent both sides.
Distribution: NE, N and NW China, Shandong, Jiangsu and Henan; Mongolia, Korea, Japan, Siberia
Habitat: Slopes and roadsides
Use: Forage; medicine

线叶蒿
Artemisia subulata Nakai
菊科
Asteraceae (Compositae)

【特征】多年生草本，高30～60cm；茎单生或少数，常带紫色或褐色；叶条状披针形或条形，全缘，稀具1～2齿，反卷，上面绿色，无毛，下面密被灰白色蛛丝状毛；头状花序直径2～3mm；总苞片密被蛛丝状毛。花期8～9月。
【分布】东北、华北；朝鲜、日本、俄罗斯（远东）。
【生境】山地、河谷、林缘、草甸、田埂、村舍、路旁。
【用途】饲用。

Threadleaf sage
Aster family
Perennial herb 30～60 cm tall; stems single or few, usually purplish or brownish; leaves linear-lanceolate or linear, entire, rarely 1- to 2-thooted, revolute, green and glabrous above, densely grey-white arachnoid-hairy below; heads 2～3 mm across; involucral bracts densely arachnoid-hairy. Flowering August to September.
Distribution: NE and N China; Korea, Japan, Russia (Far East)
Habitat: Mountains and valleys, forest margins, meadows, field ridges, villages and roadsides
Use: Forage

裂叶蒿
Artemisia tanacetifolia L.
菊科
Asteraceae (Compositae)

【特征】多年生草本，高20～75cm；茎直立，单一或少数，中部以上有分枝；下部叶2～3回栉齿状羽状全裂至深裂，上部叶1～2回栉齿状羽状全裂；头状花序球形或半球形，直径2～3mm，下垂；总苞片初时无毛或被短柔毛，后变无毛。花期7～8月。

【分布】东北、华北、陕西、宁夏、甘肃；蒙古、朝鲜、西伯利亚、中亚、欧洲、阿拉斯加。

【生境】草甸、草原、林缘、灌丛。

【用途】饲用。

Tansyleaf sage
Aster family
Perennial herb 20～75 cm tall; stems erect, single or few, branched in half above; lower leaves pectinate-pinnately divided to parted twice to thrice, upper leaves pectinate-pinnately divided once to twice; heads globose or hemispheric, 2～3 mm across, nodding; involucral bracts glabrous or pubescent to glabrate. Flowering July to August.
Distribution: NE and N China, Shaanxi, Ningxia and Gansu; Mongolia, Korea, Siberia, Central Asia, Europe, Alaska
Habitat: Meadows, steppe, forest borders and thickets
Use: Forage

羽叶鬼针草
Bidens maximowicziana Oett.
菊科
Asteraceae (Compositae)

【特征】一年生草本，高30～80cm；茎直立；中部叶羽状全裂，侧生裂片条形或条状披针形，边缘具锯齿；头状花序直径1～2cm，单生；外层总苞片条状披针形，叶状，边缘具疏齿和缘毛，内层者膜质，有褐色条纹；管状小花黄色，4裂。花期8～9月。

【分布】东北，内蒙古东部；朝鲜、日本、俄罗斯（西伯利亚、远东）。

【生境】河边湿地、路旁。

【用途】药用。

Featherleaf beggarticks
Aster family
Annual herb 30～80 cm tall; stems erect; middle leaves pinnatisect, lateral segments linear or linear-lanceolate, serrate; heads 1～2 cm across, solitary; outer involucral bracts linear-lanceolate, leaf-like, margins sparsely toothed and ciliate, inner ones membranous with brown streaks; disk florets yellow, 4-lobed. Flowering August to September.
Distribution: NE China, E Inner Mongolia; Korea, Japan, Russia (Siberia and Far East)
Habitat: Wetlands along riversides, roadsides
Use: Medicine

星毛短舌菊
Brachanthemum pulvinatum (Hand.-Mazz.) Shih
菊科
Asteraceae (Compositae)

【特征】半灌木，高10~30cm；茎自基部多分枝；叶羽状或近掌状3~5深裂，灰绿色，密被星状毛；头状花序单生枝端，半球形；总苞片卵形，边缘宽膜质，外层者被星状毛；舌状小花和管状小花黄色。花期8~9月。
【分布】我国西北特有。
【生境】戈壁荒漠砾石质坡地和覆沙地。
【用途】饲用。

Cushion brachanthemum
Aster family
Subshrub 10~30 cm tall; stems much branched from base; leaves pinnately or subpalmately 3- to 5-parted, grey-green, densely stelipilous; head solitary and terminal, hemispherical; involucral bracts ovate, with widely membranous edges, outer ones stelipilous; both ray florets and disk florets yellow. Flowering August to September.
Distribution: Endemic to NW China
Habitat: Gravelly slopes and sanded sites in gobi-desert
Use: Forage

短喙粉苞菊
Chondrilla brevirostris Fisch. et C.A. Mey.
菊科
Asteraceae (Compositae)

【特征】多年生草本，高30～100cm，含白色乳汁；茎直立，多分枝；茎生叶条形或丝状；头状花序具小花9～12枚；总苞圆柱形，总苞片2层；舌状小花黄色；瘦果长4～5mm，喙长约0.6mm。花期6～9月。
【分布】新疆；哈萨克斯坦、俄罗斯。
【生境】沙地、草甸、田边。
【用途】饲用。

Shortbeak skeletonweed
Aster family
Perennial herb 30～100 cm tall, with white milky juice; stems erect, much branched; cauline leaves linear or filiform; heads with 9～12 florets; involucre cylindric, involucral bracts in 2 series; ligulate florets yellow; achenes 4～5 mm long with a beak about 0.6 mm. Flowering June to September.
Distribution: Xinjiang; Kazakhstan, Russia
Habitat: Sands, meadows and farmland sides
Use: Forage

北疆粉苞苣
Chondrilla lejosperma Kar. et Kir.
菊科
Asteraceae (Compositae)

【特征】多年生草本，高30～120cm，含白色乳汁；茎直立，自基部多分枝；茎下部叶矩圆形或披针形，上部叶条形至倒披针形，带蓝灰色；头状花序具小花9～11枚；总苞圆柱形，总苞片2层；舌状小花黄色；瘦果长3～5mm，喙长1.3～3mm。花期5～9月。
【分布】新疆；中亚、蒙古。
【生境】石质山坡。
【用途】饲用。

North Xinjiang skeletonweed
Aster family
Perennial herb 30～120 cm tall, with white milky juice; stems erect, much branched from base; lower cauline leaves oblong or lanceolate, upper leaves linear to oblanceolate, bluish-grey; heads with 9～11 florets; involucre cylindric, involucral bracts in 2 series; ligulate florets yellow; achenes 3～5 mm long with a beak to 1.3～3 mm long. Flowering May to September.
Distribution: Xinjiang; Central Asia, Mongolia
Habitat: Rocky slopes
Use: Forage

还阳参
Crepis crocea (Lam.) Babc.
菊科
Asteraceae (Compositae)

【特征】多年生草本，高5～30cm，含白色乳汁；茎直立；基生叶丛生，倒披针形，边缘具波状齿或倒向锯齿至羽状半裂，茎上部叶披针形或条形，全缘或羽状分裂；头状花序单生于枝端，或少数排列成疏伞房状；舌状小花黄色。花期6～7月。
【分布】东北、华北，西藏；蒙古、俄罗斯（西伯利亚、远东）。
【生境】草原至荒漠草原带沙砾质山坡、黄土丘陵、田边、路旁。
【用途】饲用；药用。

Common hawksbeard
Aster family
Perennial herb 5～30 cm tall, with white milky juice; stems erect; basal leaves tufted, oblanceolate, margins repand-toothed or retrorse-serrate to pinnatifid, upper leaves lanceolate or linear, entire or pinnatifid; heads solitary and terminal, or few loosely corymbiform-arranged; ligulate florets yellow. Flowering June to July.
Distribution: NE and N China, Tibet; Mongolia, Russia (Siberia and Far East)
Habitat: Sandy and gravelly slopes, loess hills, farmland sides and roadsides in steppe to desert-steppe zones
Use: Forage; medicine

拍摄人：赵利清 Photos by Zhao Liqing

拍摄人：赵 凡 Photos by Zhao Fan

山柳菊
Hieracium umbellatum L.
菊科
Asteraceae (Compositae)

- 【特征】多年生草本，高30～100cm，含白色乳汁；茎直立，单生或少数，不分枝；茎生叶披针形至条形，全缘或有疏齿；头状花序少数或多数，排列成伞房状或圆锥状；总苞片黑绿色；舌状小花黄色，下部有长柔毛。花期7～8月。
- 【分布】东北、华北、西北、华中、西南；蒙古、日本、俄罗斯、印度、巴基斯坦、中亚、伊朗、欧洲。
- 【生境】林下、林缘、草甸。
- 【用途】饲用；染制羊毛和丝绸。

Narrowleaf hawkweed (Northern hawkweed)
Aster family
Perennial herb 30～100 cm tall, with white milky juice; stems erect, single or few, unbranched; cauline leaves lanceolate to linear, entire or spaced-serrate; heads few or many, corymbiform- or paniculate-arranged; involucral bracts black-green; ligulate florets yellow, pilose the lower portion. Flowering July to August.
Distribution: NE, N, NW, C and SW China; Mongolia, Japan, Russia, India, Pakistan, Central Asia, Iran, Europe
Habitat: Forests, forest margins and meadows
Use: Forage; dyeing wool and silk

贺兰女蒿
Hippolytia alashanensis (Ling) Shih
菊科
Asteraceae (Compositae)

【特征】小半灌木，高约30cm；茎粗壮，多分枝；叶羽状深裂或浅裂，上面绿色，下面灰白色，密被短柔毛，上部叶渐小，全缘或3浅裂；头状花序钟状，4~8枚伞房状排列；外层总苞片边缘褐色，膜质，背部被短柔毛；管状小花黄色，外面有腺点。花期7~9月。
【分布】内蒙古（贺兰山）、宁夏、甘肃。
【生境】向阳石质山坡和石缝。
【用途】饲用。

Alashan hippolytia
Aster family
Shrublet about 30 cm tall; stems stout, much branched; leaves pinnately parted or lobed, green above, grey-white and densely pubescent below, upper leaves reduced, entire or 3-lobed; heads campanulate, 4~8 corymbiform-arranged; outer involucral bracts with brown-membranous edges and pubescent back; tubular florets yellow, glandular outside. Flowering July to September.
Distribution: Inner Mongolia (Helan Mountains), Ningxia and Gansu
Habitat: Sunny rocky slopes and rock crevices
Use: Forage

拍摄人：张洪江 Photos by Zhang Hongjiang

总状土木香
Inula racemosa Hook. f.
菊科
Asteraceae (Compositae)

【特征】多年生草本，高60～200cm；茎直立，有分枝；叶大，椭圆状披针形至卵状披针形，边缘有锯齿或重齿，下面被茸毛；头状花序直径5～8cm，总状排列，梗长0.5～4cm；外层总苞片宽大，被茸毛，内层总苞片干膜质；舌状小花和管状小花黄色。花期8～9月。
【分布】新疆；印度。
【生境】草原地带湿润草地和水边。
【用途】饲用；药用。

Clustered yellowhead
Aster family
Perennial herb 60～200 cm tall; stems erect, branched; leaves large, elliptic-lanceolate to ovate-lanceolate, serrate or duplicato-serrate, downy below; heads 5～8 cm across, racemose-arranged, pedicels 0.5～4 cm long; outer involucral bracts wide and large, downy, inner ones scarious; both ray florets and disk florets yellow. Flowering August to September.
Distribution: Xinjiang; India
Habitat: Moist meadows and watersides in steppe zone
Use: Forage; medicine

山苦荬
Ixeris chinensis (Thunb.) Nakai
菊科
Asteraceae (Compositae)

【特征】多年生草本，高10～30cm，含白色乳汁；茎直立或斜升，有时斜倚；基生叶莲座状，条形至条状披针形，全缘或具疏齿，或羽状浅裂至深裂，灰绿色，茎生叶少数；头状花序排列成稀疏的伞房状；舌状小花黄色、白色或带紫色。花期6～7月。
【分布】东北、华北、华东、华南、西南；朝鲜、日本、俄罗斯（西伯利亚、远东）、越南。
【生境】田野、田间、路旁。
【用途】饲用；食用；药用。

Chinese ixeris
Aster family
Perennial herb 10～30 cm tall, with white milky juice; stems erect or ascending, sometimes decumbent; basal leaves rosulate, linear to linear-lanceolate, entire or spaced-serrate, or pinnately lobed to parted, grey-green, cauline leaves few; heads loosely corymbose-arranged; ligulate florets yellow, white or purplish. Flowering June to July.
Distribution: NE, N, E, S and SW China; Korea, Japan, Russia (Siberia and Far East), Viet Nam
Habitat: Fields, farmlands and roadsides
Use: Forage; edible; medicine

抱茎苦荬菜
Ixeris sonchifolia (Bunge) Hance
菊科
Asteraceae (Compositae)

【特征】多年生草本,高30~50cm,含白色乳汁;茎直立,上部有分枝;基生叶多数,矩圆形,边缘具齿或羽状深裂,茎生叶较小,基部扩大成耳形或戟形,抱茎;头状花序伞房状排列;舌状小花黄色。花期6~7月。
【分布】东北、华北;朝鲜、俄罗斯(远东)。
【生境】山坡草地、田野、撂荒地、路旁。
【用途】饲用;药用。

Clasping ixeris
Aster family
Perennial herb 30~50 cm tall, with white milky juice; stems erect, branched above; basal leaves many, oblong, margins toothed or pinnatipartite, cauline leaves smaller, basally enlarged to auriculate or hastate, clasping; heads corymbose-arranged; ligulate florets yellow. Flowering June to July.
Distribution: NE and N China; Korea, Russia (Far East)
Habitat: Sloping grasslands, fields, abandoned lands and roadsides
Use: Forage; medicine

团球火绒草
Leontopodium conglobatum (Turcz.) Hand.-Mazz.
菊科
Asteraceae (Compositae)

【特征】多年生草本，高10～50cm；茎直立，不分枝，被蛛丝状绵毛；叶狭倒披针形至披针状条形，被灰白色蛛丝状绵毛；苞叶多数，卵形或卵状披针形，被白色厚绵毛；头状花序5至多数密集成团球状伞房花序；总苞被白色绵毛。花期6～8月。
【分布】黑龙江北部、内蒙古东部；蒙古、西伯利亚。
【生境】草原、灌丛。
【用途】药用；观赏。

Globe edelweiss
Aster family
Perennial herb 10～50 cm tall; stems erect, unbranched, arachnoid-lanate; leaves narrowly oblanceolate to lanceolate-linear, grey-white arachnoid-lanate; bracteal leaves many, ovate or ovate-lanceolate, thickly white-lanate; heads 5 to numerous, crowded in a glomerate corymb; involucre white-lanate. Flowering June to August.
Distribution: N Heilongjiang and E Inner Mongolia; Mongolia, Siberia
Habitat: Grasslands and scrublands
Use: Medicine; ornamental

拍摄人：赵利清 Photos by Zhao Liqing

拍摄人：赵 凡 Photos by Zhao Fan

蹄叶橐吾
Ligularia fischeri (Ledeb.) Turcz.
菊科
Asteraceae (Compositae)

【特征】多年生草本，高60～120cm；茎直立；基生叶和茎下部叶具长柄，叶脉掌状，叶片肾形或心形，侧裂片近圆形，边缘具齿，茎上部叶具短柄，鞘膨大；头状花序总状排列；总苞钟状；舌状小花5～9，黄色，管状小花多数，黄色。花期7～8月。

【分布】全国大部分省区（安徽、浙江、西藏除外）；蒙古、朝鲜、俄罗斯（西伯利亚、远东）。

【生境】草甸、灌丛、林缘及林间。

【用途】观赏；根药用。

Fischer ragweed
Aster family
Perennial herb 60～120 cm tall; stems erect; basal and lower cauline leaves long-petiolate, nerves palmate, blades reniform or cordate, lateral lobes suborbicular, serrate, upper leaves short-petiolate with inflated sheath; heads racemose-arranged; involucre campanulate; ray florets 5～9, yellow, disk florets numerous, yellow. Flowering July to August.

Distribution: Almost throughout China except Anhui, Zhejiang and Tibet; Mongolia, Korea, Russia (Siberia and Far East)

Habitat: Meadows, thickets, forest margins and openings

Use: Ornamental; roots for medicine

狭苞橐吾
Ligularia intermedia Nakai
菊科
Asteraceae (Compositae)

【特征】多年生草本,高40~100cm;茎直立,上部被蛛丝状柔毛,下部光滑;基生叶和茎下部叶具长柄,叶脉掌状,叶片肾形或心形,边缘具齿,茎上部叶具短柄或无柄,鞘略膨大;头状花序总状排列;总苞筒状;舌状小花4~6,黄色,管状小花7~16,黄色。花期7~8月。
【分布】东北、华北、华中,云南、四川、贵州;朝鲜、日本。
【生境】草甸、灌丛、林缘及林间。
【用途】观赏;根药用。

Narrowbract ragwort
Aster family
Perennial herb 40~100 cm tall; stems erect, arachnoid-pubescent above, smooth below; basal and lower cauline leaves long-petiolate, nerves palmate, blades reniform or cordate, serrate, upper leaves short-petiolate or sessile, with slightly inflated sheath; heads racemose-arranged; involucre terete; ray florets 4~6, yellow, disk florets 7~16, yellow. Flowering July to August.
Distribution: NE, N, C China, Yunnan, Sichuan and Guizhou; Korea, Japan
Habitat: Meadows, thickets, forest margins and openings
Use: Ornamental; roots for medicine

拍摄人:赵利清 Photo by Zhao Liqing

全缘橐吾
Ligularia mongolica (Turcz.) DC.
菊科
Asteraceae (Compositae)

【特征】多年生草本，高30～110cm，全株灰绿色，光滑；茎直立；基生叶和茎下部叶具长柄，叶脉羽状，叶片卵形至椭圆形，全缘，茎上部叶无柄，抱茎；头状花序总状排列；总苞狭钟状或筒状；舌状小花通常3～5，黄色，管状小花5～10，黄色。花期6～7月。
【分布】东北、华北；蒙古、朝鲜、俄罗斯（远东）。
【生境】灌丛、草甸、草甸草原。
【用途】观赏；根药用。

Mongolian ragwort
Aster family
Perennial herb 30～110 cm tall, grey-green, smooth throughout; stems erect; basal and lower cauline leaves long-petiolate, nerves pinnate, blades ovate to elliptic, entire, upper leaves sessile, clasping; heads racemose-arranged; involucre narrowly campanulate or terete; ray florets usually 3～5, yellow, disk florets 5～10, yellow. Flowering June to July.
Distribution: NE and N China; Mongolia, Korea, Russia (Far East)
Habitat: Thickets, meadows and meadow-steppe
Use: Ornamental; roots for medicine

毛连菜
Picris davurica Fisch. ex Hornem.
(*Picris japonica* Thunb.)
菊科
Asteraceae (Compositae)

【特征】二年生草本，高30~80cm；茎直立，被钩状硬毛，上部有分枝；叶倒披针形至条状披针形，边缘具波状齿和钩状硬毛，两面被钩状硬毛；头状花序多数，排列成伞房圆锥状；总苞片黑绿色，被硬毛和短柔毛；舌状小花淡黄色。花期7~8月。

【分布】东北、华北、西北、华东、华中、西南；日本、俄罗斯（西伯利亚、远东）。

【生境】林缘、草甸、沟谷、路旁。

【用途】药用。

Wooly oxtongue
Aster family
Biennial herb 30~80 cm tall; stems erect, glochidiate, branched above; leaves oblanceolate to linear-lanceolate, repand-serrate and glochidiate at margins, glochidiate both sides; many heads corymbosely paniculate-arranged; involucral bracts black-green, hispid and pubescent; ligulate florets pale yellow. Flowering July to August.
Distribution: NE, N, NW, E, C and SW China; Japan, Russia (Siberia and Far East)
Habitat: Forest margins, meadows, gullies and roadsides
Use: Medicine

拍摄人：赵 凡 Photos by Zhao Fan

叉枝鸦葱
Scorzonera muriculata Chang
菊科
Asteraceae (Compositae)

【特征】半灌木状草本，高15～45cm，含白色乳汁；茎灰绿色，有白粉，自基部等叉分枝，分枝多，形成半球状株丛；叶条形或丝状条形，常反卷成钩状；头状花序单生枝顶，具3～7舌状小花，黄色；冠毛基部连合成环，整体脱落。花期5～7月。
【分布】内蒙古、甘肃。
【生境】戈壁。
【用途】饲用；药用。

Forked vipergrass
Aster family
Subfrutical herb 15～45 cm tall, with white milky juice; stems grey-green, farinaceous, equally dichotomous-branched from base, branches numerous, forming a hemispheric stand; leaves linear or filiform-linear, usually reflexed to hamate; head solitary and terminal, with ligulate florets 3～7, yellow; pappus basally connected into a ring, falling off together. Flowering May to July.
Distribution: Inner Mongolia and Gansu
Habitat: Gobi
Use: Forage; medicine

纤细绢蒿
Seriphidium gracilescens (Krasch. et Iljin) Poljak.
菊科
Asteraceae (Compositae)

【特征】半灌木状草本，高15～30cm；茎多数，密丛生；叶质稍厚，下部叶2至3回羽状全裂，上部叶羽状全裂或不裂，两面密被绒毛，并有腺点；头状花序直径1.5～2.5mm，椭圆状或矩圆状，直立；外层总苞片边缘膜质；管状小花黄色。花期8～9月。
【分布】新疆；蒙古、西伯利亚。
【生境】海拔500～3 000m的荒漠草原、河谷阶地。
【用途】饲用。

Slender sagebrush
Aster family
Subfrutical herb 15～30 cm tall; stems many, strongly caespitose; leaves slightly thick, lower leaves pinnatisect twice to thrice, upper leaves pinnatisect or undivided, densely tomentose with glands; heads 1.5～2.5 mm across, ellipsoid to oblong, erect; outer involucral bracts with membranous edges; tubular florets yellow. Flowering August to September.
Distribution: Xinjiang; Mongolia, Siberia
Habitat: Desert-steppe and valley terraces at 500～3 000 m
Use: Forage

白茎绢蒿
Seriphidium terrae-albae
(Krasch.) Poljak.
菊科
Asteraceae (Compositae)

【特征】半灌木状草本，高20～35cm；茎多数，密丛生；叶2回羽状全裂，上部叶不裂，两面被蛛丝状绒毛；头状花序直径1～2mm，卵状或长卵状，直立或斜展；外层总苞片背部突起，密被蛛丝状柔毛；管状小花黄色。花期8～9月。
【分布】新疆；中亚、蒙古。
【生境】海拔500～2 000m的沙漠、戈壁、半固定沙丘。
【用途】饲用。

Whitestem sagebrush
Aster family
Subfrutical herb 20～35 cm tall; stems many, strongly caespitose; leaves bipinnatisect, upper leaves undivided, arachnoid-tomentose; heads 1～2 mm across, ovoid to long-ovoid, erect or ascending; outer involucral bracts prominent and densely arachnoid-pubescent on back; tubular florets yellow. Flowering August to September.
Distribution: Xinjiang; Central Asia, Mongolia
Habitat: Sandy desert, gobi and semi-fixed dunes at 500～2 000 m
Use: Forage

苣荬菜
Sonchus arvensis L.
菊科
Asteraceae (Compositae)

【特征】多年生草本,高20~80cm,含白色乳汁;茎直立;基生叶与茎下部叶羽状浅裂或具波状疏齿,中部叶多少呈耳状抱茎,上部叶渐小;头状花序伞房状排列或单生;总苞片被短柔毛;舌状小花黄色。花期6~8月。
【分布】东北、华北、西北;蒙古、朝鲜、日本、俄罗斯(远东)。
【生境】草甸、田间、村舍、路旁。
【用途】饲用;食用;药用。

Field sow thistle
Aster family
Perennial herb 20~80 cm tall, with white milky juice; stems erect; basal and lower cauline leaves pinnatilobate or sparsely undulate-toothed, middle leaves somewhat auriculate-clasping, upper leaves reduced; heads corymbose-arranged or solitary; involucral bracts pubescent; ligulate florets yellow. Flowering June to August.
Distribution: NE, N and NW China; Mongolia, Korea, Japan, Russia (Far East)
Habitat: Meadows, farmlands, villages and roadsides
Use: Forage; edible; medicine

拍摄人：卢欣石 Photos by Lu Xinshi

苦苣菜
***Sonchus oleraceus* L.**
菊科
Asteraceae (Compositae)

【特征】一年生或二年生草本，高30～80cm，含白色乳汁；茎直立；叶羽状深裂、半裂或大头羽裂，边缘具刺齿，下部叶具柄，柄基扩大抱茎，上部叶无柄，基部戟状耳形，抱茎；头状花序伞房状排列；总苞片疏生腺毛；舌状小花黄色。花期6～8月。
【分布】全国各地；世界广布。
【生境】田野、村舍、路旁。
【用途】饲用；食用；药用。

Common sow thistle
Aster family
Annual or biennial herb 30～80 cm tall, with white milky juice; stems erect; leaves pinnately parted, cleft or lyrate, margins spinose-toothed, lower leaves with petiole basally enlarged and clasping, upper leaves sessile, basally hastate-auriculate and clasping; heads corymbose-arranged; involucral bracts sparsely glandular-hairy; ligulate florets yellow. Flowering June to August.
Distribution: Throughout China; cosmopolitan
Habitat: Fields, villages and roadsides
Use: Forage; edible; medicine

亚洲蒲公英
Taraxacum asiaticum Dahlst.
菊科
Asteraceae (Compositae)

【特征】多年生草本，高10～30cm；含白色乳汁；无茎；叶羽状深裂或浅裂，顶端裂片戟形或狭戟形，侧裂片平展或向下，裂片远隔，其间夹生小裂片或小齿；花葶数个；总苞片先端有不明显的角状突起；舌状小花淡黄色或白色。花期5～7月。

【分布】东北、华北、西北、四川、湖北；蒙古、西伯利亚、中亚。

【生境】草甸、河滩、林缘、村舍附近。

【用途】饲用；药用。

Asian dandelion
Aster family
Perennial herb 10～30 cm tall, with white milky juice; acaulescent; leaves pinnately parted or lobed, terminal segment hastate or narrowly hastate, lateral segments explanate or directed downward and spaced, intersegment lobulate or toothed; scapes several; involucral bracts inconspicuously cornute at apex; ligulate florets pale yellow or white. Flowering May to July.
Distribution: NE, N and NW China, Sichuan and Hubei; Mongolia, Siberia, Central Asia
Habitat: Meadows, flood lands, forest edges and village sides
Use: Forage; medicine

多裂蒲公英
Taraxacum dissectum (Ledeb.) Ledeb.
菊科
Asteraceae (Compositae)

【特征】多年生草本，高5～25cm；含白色乳汁；无茎；叶羽状全裂，顶端裂片长三角状戟形，侧裂片向下，条形，全缘，裂片间无齿或小裂片，两面被蛛丝状毛；花葶1～6个；总苞片绿色，先端带紫色，无角；舌状小花黄色。花期6～8月。
【分布】西北，内蒙古、山西、陕西、西藏；蒙古、西伯利亚。
【生境】盐渍化草甸和砾质砂地。
【用途】饲用。

Cutleaf dandelion
Aster family
Perennial herb 5～25 cm tall, with white milky juice; aculescent; leaves pinnatisect, terminal segment long triangular-hastate, lateral segments directed downward, linear, entire, intersegment neither toothed nor lobulate, arachnoid-hairy; scapes 1～6; involucral bracts green with purplish apex, cornet without; ligulate florets yellow. Flowering June to August.
Distribution: NW China, Inner Mongolia, Shanxi, Shaanxi and Tibet; Mongolia, Siberia
Habitat: Saline meadows and gravelly sands
Use: Forage

蒲公英
Taraxacum mongolicum Hand.-Mazz.
菊科
Asteraceae (Compositae)

【特征】多年生草本，高10～30cm；含白色乳汁；无茎；叶大头羽裂或倒向羽裂，有时为羽状浅裂或不分裂而具波状齿；花葶数个；总苞片先端有明显的角状突起；舌状小花黄色。花期5～7月。
【分布】东北、华北、西北、华东、华中、西南；蒙古、朝鲜、俄罗斯。
【生境】草地、田野、河滩、路边。
【用途】饲用；药用。

Mongolian dandelion
Aster family
Perennial herb 10～30 cm tall, with white milky juice; acaulescent; leaves lyrate or runcinate, sometimes pinnatilobate or undivided with undulate teeth; scapes several; involucral bracts conspicuously cornute at apex; ligulate florets yellow. Flowering May to July.
Distribution: NE, N, NW, E, C and SW China; Mongolia, Korea, Russia
Habitat: Grasslands, fields, flood lands and roadsides
Use: Forage; medicine

拍摄人：张洪江 Photos by Zhang Hongjiang

草地婆罗门参
Tragopogon pratensis L.
菊科
Asteraceae (Compositae)

【特征】二年生草本，高25～100cm，含白色乳汁；茎直立，单一，无毛；叶条形至条状披针形，基部扩大抱茎，全缘或边缘皱波状，上部叶渐小；头状花序单生茎顶，或少数排列成伞房状；舌状小花黄色；瘦果喙等长或长于果体。花期6～7月。
【分布】新疆；西伯利亚、高加索、欧洲、北美洲。
【生境】山坡草地、林缘、灌丛。
【用途】饲用。

Meadow salsify
Aster family
Biennial herb 25～100 cm tall, with white milky juice; stems erect, single, glabrous; leaves linear to linear-lanceolate, basally enlarged and clasping, entire or repand, upper leaves smaller; heads solitary and terminal, or few corymbose-arranged; ligulate florets yellow; achenes with a beak equaling or longer than the body. Flowering June to July.
Distribution: Xinjiang; Siberia, Caucasia, Europe, North America
Habitat: Sloping grasslands, forest margins and thickets
Use: Forage

碱黄鹌菜
Youngia stenoma (Turcz.) Ledeb.
菊科
Asteraceae (Compositae)

【特征】多年生草本，高10～50cm，含白色乳汁；茎直立，不分枝；叶质厚，灰绿色，条形或条状披针形，全缘或边缘有浅波状齿；头状花序总状或狭圆锥状排列；总苞片背部近顶端有角状突起；舌状小花8～12，黄色。花期7～8月。
【分布】东北，内蒙古、甘肃、西藏；西伯利亚。
【生境】盐化草甸、草原沙地。
【用途】饲用；药用。

Alkali false hawksbeard
Aster family
Perennial herb 10～50 cm tall, with white milky juice; stems erect, unbranched; leaves thick, grey-green, linear or linear-lanceolate, entire or repand-toothed; heads racemose-arranged or narrowly paniculate-arranged; involucral bracts cornute near back top; ligulate florets 8～12, yellow. Flowering July to August.
Distribution: NE China, Inner Mongolia, Gansu and Tibet; Siberia
Habitat: Saline meadows, and sands in steppe
Use: Forage; medicine

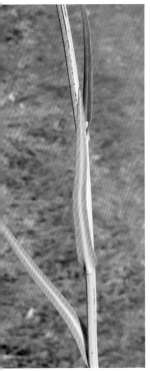

拍摄人：赵利清 Photo by Zhao Liqing

少花顶冰花
***Gagea pauciflora* (Turcz. ex Trautv.) Ledeb.**
百合科
Liliaceae

【特征】多年生草本，高20～70cm；基生叶1枚，茎生叶2～5枚，披针状条形；花序近总状，具花1～3枚；花被片6，绿黄色；花药长2～3.5mm；柱头3深裂。花期5～6月。
【分布】河北、内蒙古、陕西、甘肃、青海、西藏；蒙古、俄罗斯（西伯利亚、远东）。
【生境】山地草甸和灌丛。
【用途】饲用。

Fewflower gagea
Lily family
Perennial herb 20～70 cm tall; basal leaf 1, cauline leaves 2～5, lanceolate-linear; inflorescence nearly racemose, with 1～3 flowers; tepals 6, green-yellow; anthers 2～3.5 mm long; stigma 3-parted. Flowering May to June.
Distribution: Hebei, Inner Mongolia, Shaanxi, Gansu, Qinghai and Tibet; Mongolia, Russia (Siberia and Far East)
Habitat: Montane meadows and thickets
Use: Forage

北黄花菜
Hemerocallis lilioasphodelus L.
百合科
Liliaceae

【特征】多年生草本；叶基生，条形，长20～70cm；花序分枝，具4至多枚花；花梗不等长；花被黄色，花被管长1.5～2.5cm，花被裂片6，长5～7cm。花期6～7月。
【分布】东北、华北，山东、陕西、甘肃；俄罗斯、欧洲。
【生境】草甸、灌丛。
【用途】饲用；食用；观赏；药用。

Yellow daylily (Lemon lily)
Lily family
Perennial herb; leaves basal, linear, 20～70 cm long; inflorescence branched, with 4 to more flowers; pedicels unequal; perianth yellow, tube 1.5～2.5 cm long, perianth with 6 segments, 5～7 cm long. Flowering June to July.
Distribution: NE, N China, Shandong, Shaanxi and Gansu; Russia, Europe
Habitat: Meadows and thickets
Use: Forage; edible; ornamental; medicine

拍摄人：赵 凡 Photo by Zhao Fan

单脉大黄
***Rheum uninerve* Maxim.**
蓼科
Polygonaceae

【特征】多年生草本，高10～30cm；无茎；叶基生，叶片纸质，卵形，基部近圆形或宽楔形，边缘弱皱波状，叶脉掌羽状；圆锥花序2～5，花2～4枚簇生；花被片6，粉色；瘦果具3棱，沿棱具宽翅，翅膜质，粉紫色。花期5～7月。
【分布】内蒙古西部、宁夏、甘肃。
【生境】荒漠草原至荒漠地区的石质山坡、岩石缝、冲刷沟。
【用途】饲用。

Monovein rhubarb
Buckwheat family
Perennial herb 10～30 cm tall; aculescent; leaves basal, blades papery, ovate, base suborbicular or broadly cuneate, margins sinuolate, veins palmately pinnate; panicles 2～5, with flowers in fascicles of 2～4; tepals 6, pink; achenes trigonous, widely winged on acies, wings membranous and pinkish-purple. Flowering May to July.
Distribution: W Inner Mongolia, Ningxia and Gansu
Habitat: Rocky slopes, rock crevices and water-eroded ditches in steppe-desert to desert regions
Use: Forage

头状石头花 (头花丝石竹)
Gypsophila capituliflora Rupr.
石竹科
Caryophyllaceae

【特征】多年生垫状草本，高10～30cm；茎多数，基部叶密集；叶近三棱状条形；花多数，密集成头状聚伞花序；花瓣5，粉色或淡紫色。花期7～9月。
【分布】内蒙古中部和西部、宁夏、新疆；蒙古、阿富汗、中亚。
【生境】石质山坡和山顶石缝。
【用途】饲用。

Crested baby's breath
Pink family
Perennial cushion-like herb 10～30 cm tall; stems many, with dense leaves at base; leaves nearly trigonous-linear; flowers numerous, crowded in a capitate cyme; petals 5, pink or pale purple. Flowering July to September.
Distribution: C and W Inner Mongolia, Ningxia and Xinjiang; Mongolia, Afghanistan, Central Asia
Habitat: Rocky slopes and rock crevices on hilltops
Use: Forage

女娄菜
Melandrium apricum (Turcz.) Rohrb.
(*Silene aprica* Turcz.)
石竹科
Caryophyllaceae

【特征】一年生或二年生草本，高20~70cm，全株密被倒向短柔毛；茎直立；叶条状披针形或披针形；聚伞花序顶生或腋生；花萼卵状钟形，果期膨大，密被短柔毛；花瓣5，白色或粉色，与萼等长或稍长，2裂；蒴果6齿裂。花期5~7月。

【分布】东北、华北、西北、西南、华东；蒙古、朝鲜、日本、俄罗斯（西伯利亚、远东）

【生境】平原、丘陵、山地，疏林、草地。

【用途】饲用；药用。

拍摄人：赵利清 Photo by Zhao Liqing

Sunward catchfly
Pink family
Annual or biennial herb 20~70 cm tall, densely retrorse-pubescent throughout; stems erect; leaves linear-lanceolate or lanceolate; cymes terminal or axillary; calyx ovoid-campanulate, inflated in fruit, densely pubescent; petals 5, white or pink, equal to or slightly longer than the calyx, bifid; capsules 6-toothed. Flowering May to July.
Distribution: NE, N, NW, SW and E China; Mongolia, Korea, Japan, Russia (Siberia and Far East)
Habitat: Plains, hills and mountains, open woodlands and grasslands
Use: Forage; medicine

兴安女娄菜
Melandrium brachypetalum (Fisch. ex Hornem.) Fenzl
(*Lychnis brachypetala* Fisch. ex Hornem.)
石竹科
Caryophyllaceae

【特征】多年生草本，高20~50cm；茎直立，密被腺质柔毛；叶条状倒披针形至披针形，密被短柔毛；聚伞状圆锥花序，花少数；花梗和花萼密被腺毛；花萼圆筒形，果期膨大；花瓣5，粉红色至紫红色，与萼等长或稍长，2裂；蒴果10齿裂。花期6~7月。
【分布】东北、西北，内蒙古；蒙古、俄罗斯。
【生境】山地林缘和草甸。
【用途】饲用。

Shortpetal campion
Pink family
Perennial herb 20~50 cm tall; stems erect, densely glandular-pubescent; leaves linear-oblanceolate to lanceolate, densely pubescent; thyrse with few flowers; pedicel and calyx densely glandular-hairy; calyx tubular, inflated in fruit; petals 5, pink to purplish-red, equal to or slightly longer than the calyx, bifid; capsules 10-toothed. Flowering June to July.
Distribution: NE and NW China, Inner Mongolia; Mongolia, Russia
Habitat: Montane forest margins and meadows
Use: Forage

拍摄人：赵利清 Photo by Zhao Liqing

拍摄人：赵利清 Photo by Zhao Liqing

华北八宝
Hylotelephium tatarinowii (Maxim.) H. Ohba
景天科
Crassulaceae

【特征】多年生肉质草本，高7～15cm，茎多数，直立或倾斜，不分枝；叶互生，条状倒披针形至倒披针形，边缘有疏齿至浅裂；伞房状聚伞花序顶生，花密生；花瓣5，粉红色；花药紫色；蓇葖果。花期7～8月。
【分布】华北。
【生境】海拔1 000～3 000m的山地石缝。
【用途】饲用。

North China stonecrop
Stonecrop family
Perennial fleshy herb 7～15 cm tall; stems many, erect or ascending, unbranched; leaves alternate, linear-oblanceolate to oblanceolate, margins sparsely toothed to lobed; corymbose cymes terminal, with dense flowers; petals 5, pink; anthers purple; fruit a follicle. Flowering July to August.
Distribution: N China
Habitat: Mountain rock crevices at 1 000～3 000 m
Use: Forage

高山地榆
Sanguisorba alpina Bunge
蔷薇科
Rosaceae

【特征】多年生草本，高30～80cm；茎通常分枝；单数羽状复叶，小叶片椭圆形或长椭圆形，基部截形或微心形，边缘有缺刻状锯齿；穗状花序粗大，下垂；花由基部向上逐渐开放；花萼白色带粉色，萼片4；无花瓣；花丝比萼片长2～3倍。花期7～8月。

【分布】内蒙古（贺兰山）、宁夏、甘肃、新疆；蒙古、朝鲜、西伯利亚、中亚。

【生境】海拔1 200～2 800m的草甸、沼泽、林缘、溪谷。

【用途】饲用；药用；提制栲胶。

Alpine burnet
Rose family
Perennial herb, 30～80 cm tall; stems usually branched; leaves odd-pinnate, leaflet blades elliptic or long-elliptic, base cuneate or slightly cordate, margins incised-serrate; spikes thick, drooping; flowers gradually opening upward; calyx pinkish-white, sepals 4; petals without; filaments twice to thrice longer than the sepals. Flowering July to August.

Distribution: Inner Mongolia (Helan Mountains), Ningxia, Gansu and Xinjiang; Mongolia, Korea, Siberia, Central Asia

Habitat: Meadows, swamps, forest edges and gullies at 1 200～2 800 m

Use: Forage; medicine; extracting tannin

下图拍摄人：赵利清 Lower photo by Zhao Liqing

拍摄人：赵利清 Photos by Zhao Liqing

鬼箭锦鸡儿
Caragana jubata (Pall.) Poir.
豆科
Fabaceae (Leguminosae)

【特征】灌木，高30～200cm；茎直立或横卧，基部多分枝；偶数羽状复叶，小叶4～6对，托叶先端刚毛状，叶轴针刺状，小叶片两面被长柔毛；花单生；蝶形花冠黄白色、粉色至淡红紫色，翼瓣具单耳；荚果圆筒状，密被长柔毛。花期6～7月。
【分布】华北、西北、西南；蒙古、西伯利亚、中亚、尼泊尔、锡金、不丹。
【生境】海拔2 400～4 700m的山地至高山灌丛和林缘。
【用途】饲用；纤维材料；药用。

Shagspine peashrub
Pea family
Shrub 30～200 cm tall; stems erect or recumbent, much branched from base; leaves even-pinnate, leaflets in 4～6 pairs, stipules setiform at apex, rachis spine-like, leaflet blades villous; flowers solitary; papilionaceous corolla yellowish-white, pink to pale reddish-purple, wings 1-eared; pods cylindric, densely villous. Flowering June to July.
Distribution: N, NW and SW China; Mongolia, Siberia, Central Asia, Nepal, Sikkim, Bhutan
Habitat: Montane to alpine scrublands and forest borders at 2 400～4 700 m
Use: Forage; fiber materials; medicine

短穗柽柳
Tamarix laxa Willd.
柽柳科
Tamaricaceae

【特征】灌木，高1~2m；幼枝短粗，质脆；叶披针形、卵状披针形或菱形；总状花序短粗，长1~4cm，粗5~10mm；春季花4基数，秋季花5基数；花瓣粉红色，稀白色，开张。花期4~5月，偶尔秋季开花。
【分布】西北，内蒙古西部；蒙古、中亚、阿富汗、伊朗。
【生境】荒漠地区的盐渍低地和沙丘边缘。
【用途】盐地改良和固沙。

Shortspike tamarisk
Tamarix family
Shrub 1~2 m tall; young branchlets short and thick, fragile; leaves lanceolate, ovate-lanceolate or rhombic; racemes short and thick, 1~4 cm long, 5~10 mm thick; vernal flowers 4-merous, autumnal flowers 5-merous; petals pink, rarely white, patent. Flowering April to May, rarely autumn.
Distribution: NW China, W Inner Mongolia; Mongolia, Central Asia, Afghanistan, Iran
Habitat: Saline lowlands and dune margins in desert areas
Use: Improving saline sites and fixing dunes

拍摄人：赵利清 Photos by Zhao Liqing

细穗柽柳
Tamarix leptostachya Bunge
柽柳科
Tamaricaceae

【特征】灌木，高1～3m；叶狭卵形或卵状披针形；总状花序细长，长5～15cm，粗2～4mm；花5基数；花瓣淡紫红色、淡蓝紫色或粉红色，开张。花期5～6月。
【分布】西北，内蒙古西部；蒙古、中亚、阿富汗、巴基斯坦。
【生境】荒漠地区的湿润盐渍化渠畔和路旁。
【用途】盐地改良和水土保持。

Thinspike tamarisk
Tamarix family
Shrub 1～3 m tall; leaves narrowly ovate or ovate-lanceolate; racemes slender, 5～15 cm long, 2～4 mm thick; flowers 5-merous; petals pale purplish-red, bluish-purple or pink, patent. Flowering May to June.
Distribution: NW China, W Inner Mongolia; Mongolia, Central Asia, Afghanistan, Pakistan
Habitat: Moist saline ditch sides and roadsides in desert zone
Use: Improving saline sites, and soil conservation

沼生柳叶菜
Epilobium palustre L.
柳叶菜科
Onagraceae

【特征】多年生草本，高20~50cm；茎直立，基部具匍匐枝；叶披针形或长椭圆形，全缘，边缘内卷；花单生于叶腋；花瓣4，粉色，顶端2裂；柱头头状；蒴果条形，长3~6cm。花期7~8月。
【分布】东北、华北、西北；亚洲、欧洲、北美洲。
【生境】沼泽、草甸、水沟、溪边。
【用途】药用。

Marsh willowherb
Evening primrose family
Perennial herb 20~50 cm tall; stems erect, stoloniferous at base; leaves lanceolate to long-elliptic, entire, margins involute; flowers solitary in leaf axils; petals 4, pink, apically bifid; stigmas capitate; capsules linear, 3~6 cm long. Flowering July to August.
Distribution: NE, N and NW China; Asia, Europe, North America
Habitat: Swamps, meadows, ditches and streamsides
Use: Medicine

葛缕子
Carum carvi L.
伞形科
Apiaceae (Umbelliferae)

【特征】二年生或多年生草本，高25～70cm；茎直立，上部分枝；叶2～3回羽状全裂，末回裂片条形或披针形，茎生叶叶鞘边缘白色或粉红色宽膜质；复伞形花序，伞幅4～10；无小总苞片；花瓣5，白色或粉红色。花期6～8月。
【分布】东北、华北、西北、西南；蒙古、朝鲜、俄罗斯、伊朗、欧洲、北非、北美洲。
【生境】林缘、草甸、田边、路旁。
【用途】饲用；药用；制作食用香料；提取挥发油。

Caraway
Parsley family
Biennial or perennial herb 25～70 cm tall; stems erect, branched above; leaves pinnatisect twice to thrice, ultimate segments linear or lanceolate, sheaths of cauline leaf with white or pink and broadly membranous margins; compound umbels with 4～10 rays; bracteoles absent; petals 5, white or pink. Flowering June to August.
Distribution: NE, N, NW and SW China; Mongolia, Korea, Russia, Iran, Europe, N Africa, North America
Habitat: Forest margins, meadows, farmland sides and roadsides
Use: Forage; medicine; making edible spice; extracting volatile oil

大叶白麻
Poacynum hendersonii (Hook. f.) Woodson
(*Apocynum hendersonii* Hook. f.)
夹竹桃科
Apocynaceae

【特征】半灌木，高0.5～2.5m，具乳汁；茎直立，无毛；叶椭圆形至卵状椭圆形，边缘具细齿；圆锥状聚伞花序顶生；花冠辐状，下垂，裂片5，粉色，内面具红色脉纹，基部具副花冠；蓇葖果2，圆筒状，平行或叉生，长10～30cm。花期5～7月。
【分布】西北；蒙古、中亚。
【生境】荒漠地带的盐碱荒地、沙漠边缘、河岸、湖边、渠旁。
【用途】纤维材料；药用；蜜源；观赏；幼果可食。

Bigleaf poacynum
Dogbane family
Subshrub 0.5～2.5 m tall, with milky juice; stems erect, glabrous; leaves elliptic to ovate-elliptic, serrulate; paniculate cymes terminal; corolla rotate, nodding, 5-lobed, pink, with red vein-streaks inner side, base coronate; fruit a 2-follicle, cylindric, parallel or crossed, 10～30 cm long. Flowering May to July.
Distribution: NW China; Mongolia, Central Asia
Habitat: Saline-alkali wastelands, edges of sandy desert, river banks, lakeshores and ditch sides in desert zone
Use: Fiber materials; medicine; honey source; ornamental; young fruits edible

白麻
Poacynum pictum (Schrenk) Baill.
(*Apocynum pictum* Schrenk)
夹竹桃科
Apocynaceae

【特征】半灌木,高0.5～2m,具乳汁;茎直立,幼枝被柔毛;叶条形至条状披针形,边缘具细齿;圆锥状聚伞花序顶生;花冠辐状,下垂,裂片5,粉色,内面具红色脉纹,基部具副花冠;蓇葖果2,圆筒状,平行或叉生,长10～20cm。花期6～7月。
【分布】西北,内蒙古西部;蒙古、中亚。
【生境】荒漠地带的盐碱荒地、沙漠边缘、河岸、湖边、渠旁。
【用途】纤维材料;药用;蜜源;观赏;幼果可食。

Common poacynum
Dogbane family
Subshrub 0.5～2 m tall, with milky juice; stems erect, young branchlets pubescent; leaves linear to linear-lanceolate, serrulate; paniculate cymes terminal; corolla rotate, nodding, 5-lobed, pink, with red vein-streaks inner side, base coronate; fruit a 2-follicle, cylindric, parallel or crossed, 10～20 cm long. Flowering June to July.
Distribution: NW China, W Inner Mongolia; Mongolia, Central Asia
Habitat: Saline-alkali wastelands, edges of sandy desert, river banks, lakeshores and ditch sides in desert zone
Use: Fiber materials; medicine; honey source; ornamental; young fruits edible

无毛兔唇花
Lagochilus bungei Benth.
唇形科
Lamiaceae (Labiatae)

【特征】多年生草本，高15～30cm；茎四棱，灰白色，直立，具分枝；叶羽状深裂或顶端3裂；轮伞花序顶生，2～10花；花冠粉红色或淡粉色，外面被长柔毛，二唇形，上唇2深裂，下唇3裂。花期7～8月。
【分布】新疆；中亚。
【生境】戈壁和荒漠中的砾石质坡地。
【用途】饲用。

Altai lagochilus
Mint family
Perennial herb 15～30 cm tall; stems quadrangular, grey-white, erect, branched; leaves pinnatipartite or apically 3-lobed; verticillasters terminal, with 2～10 flowers; corolla pink or pale pink, villous outside, bilabiate, upper lip 2-parted, lower lip 3-lobed. Flowering July to August.
Distribution: Xinjiang; Central Asia
Habitat: Gobi and gravelly slopes in desert
Use: Forage

拍摄人：赵 凡 Photos by Zhao Fan

蓍 (千叶蓍)
Achillea millefolium L.
菊科
Asteraceae (Compositae)

【特征】多年生草本，高40～60cm；茎直立，被长柔毛；叶2～3回羽状全裂，末回裂片披针形，稀条形，上面密被腺点，下面被长柔毛；头状花序多数，密集成伞房状聚伞花序；舌状小花5～7，白色、粉色或淡紫红色，管状小花5齿裂。花期7～8月。

【分布】东北，内蒙古、河北、新疆；蒙古、西伯利亚、伊朗、欧洲、非洲北部、北美洲。

【生境】草甸、疏林。

【用途】饲用；药用；观赏。

Western yarrow
Aster family
Perennial herb 40～60 cm tall; stems erect, villous; leaves pinnately divided twice to thrice, ultimate segments lanceolate, rarely linear, densely glandular above, villous below; numerous heads crowded in a corymbiform cyme; ray florets 5～7, white, pink or pale purplish-red, disk florets 5-toothed. Flowering July to August.

Distribution: NE China, Inner Mongolia, Hebei and Xinjiang; Mongolia, Siberia, Iran, Europe, N Africa, North America

Habitat: Meadows and open woodlands

Use: Forage; medicine; ornamental

花蔺
Butomus umbellatus L.
花蔺科
Butomaceae

【特征】多年生水生草本；根状茎粗壮；叶基生，剑形，基部三棱形，长30～120cm；花葶直立，圆柱形，高40～70cm；苞片3；伞形花序，花多数；花梗不等长，4～10cm；花被片6，粉色；雄蕊9；心皮6，粉红色；蓇葖果具喙。花期7～8月。

【分布】东北、华北、陕西、新疆、河南、山东、湖北、江苏；广布于亚洲和欧洲。

【生境】水边、沼泽。

【用途】茎、叶编织和造纸；根茎食用。

Flowering rush
Flowering rush family
Perennial aquatic herb; rhizomes stout; leaves basal, ensiform, basally trigonous, 30～120 cm long; scapes erect, terete, 40～70 cm tall; bracts 3; umbels with many flowers; pedicels unequal, 4～10 cm long; tepals 6, pink; stamens 9; pink carpels 6; follicles beaked. Flowering July to August.
Distribution: NE and N China, Shaanxi, Xinjiang, Henan, Shandong, Hubei and Jiangsu; widespread in Asia and Europe
Habitat: Watersides and marshes
Use: Stems and leaves for weaving and papermaking; rhizomes edible

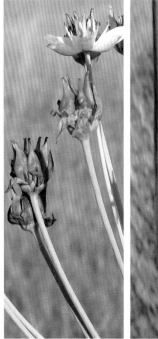

朝鲜薤
Allium sacculiferum Maxim.
百合科
Liliaceae

【特征】多年生草本；鳞茎卵球状，基部侧生小鳞茎；叶三棱状条形，中空，背面具1纵棱；花葶高70～140cm；总苞2裂；伞形花序球状；小花梗不等长，基部具小苞片；花被片6，淡紫粉色，中脉紫红色，外轮花被片舟形；花丝约为花被片的1.5倍。花期6月。
【分布】东北，内蒙古；朝鲜、日本、俄罗斯。
【生境】草甸、河边、湖边、沙地。
【用途】饲用；鳞茎食用。

Korean onion
Lily family
Perennial herb; bulbs ovoid, with lateral bulblets at base; leaves trigonous-linear, hollow, 1-ribbed below; scapes 70～140 cm tall; spathe 2-divided; umbel globose; pedicels unequal, bracteolate at base; tepals 6, pale purplish-pink, with purple-red midveins, outer tepals navicular; filaments about 1.5 times as long as the tepals. Flowering June.
Distribution: NE China, Inner Mongolia; Korea, Japan, Russia
Habitat: Meadows, riversides, lakesides and sands
Use: Forage; bulbs edible

地榆
Sanguisorba officinalis L.
蔷薇科
Rosaceae

【特征】多年生草本,高30~80cm,全株光滑无毛;茎直立,上部分枝;单数羽状复叶,小叶片卵形或矩圆状卵形,基部心形或微心形,边缘具齿;穗状花序顶生;花由顶端向下逐渐开放;花萼紫红色,萼片4;无花瓣;花丝与萼片近等长。花期7~8月。
【分布】全国各地;欧亚大陆、北美洲。
【生境】草甸、林缘、灌丛。
【用途】饲用;药用;提制栲胶;制作农药。

Great burnet
Rose family
Perennial herb 30~80 cm tall, glabrous throughout; stems erect, branched above; leaves odd-pinnate, leaflet blades ovate or oblong-ovate, base cordate or slightly cordate, margins toothed; spikes terminal; flowers gradually opening downward; calyx purple-red, sepals 4; petals without; filaments subequaling sepals. Flowering July to August.
Distribution: Throughout China; Eurasia, North America
Habitat: Meadows, forest margins and thickets
Use: Forage; medicine; extracting tannin; making pesticide

骆驼刺
Alhagi sparsifolia Shap.
豆科
Fabaceae (Leguminosae)

【特征】半灌木，高40～60cm；茎直立，多分枝，针刺长1～3.5cm；单叶，宽卵形至宽倒卵形，全缘；总状花序腋生，先端针刺状，具3～8花；蝶形花冠红色；荚果念珠状，直或稍弯。花期6～7月。
【分布】内蒙古西部、宁夏、甘肃、新疆；蒙古、中亚。
【生境】荒漠地带的沙质生境、轻度盐碱化低地。
【用途】饲用；固沙；药用；蜜源。

Camelthorn
Pea Family
Subshrub 40～60 cm tall; stems erect, much branched, thorn 1～3.5 cm long; leaves simple, broadly ovate to broadly obovate, entire; racemes axillary, terminating in a thorn, with 3～8 flowers; papilionaceous corolla red; pods moniliform, straight or slightly curved. Flowering June to July.
Distribution: W Inner Mongolia, Ningxia, Gansu and Xinjiang; Mongolia, Central Asia
Habitat: Sandy sites and lightly saline-alkali lowlands in desert zone
Use: Forage; fixing dunes; medicine; honey source

拍摄人：赵利清 Photos by Zhao Liqing

贺兰山岩黄耆
Hedysarum petrovii Yakovl.
豆科
Fabaceae (Leguminosae)

【特征】多年生草本，高4～20cm；茎多数，短缩；单数羽状复叶，小叶7～15，小叶片上面密被腺点，下面密被长柔毛；总状花序腋生，长于叶，花10～20枚；蝶形花冠红色或红紫色；荚果1～4节，节荚稍凸起，密被柔毛和硬针刺。花期6～7月。
【分布】内蒙古（贺兰山）、陕西、宁夏、甘肃。
【生境】荒漠区沟壑。
【用途】饲用；观赏。

Helan sweetvetch
Pea family
Perennial herb 4～20 cm tall; stems many, shortened; leaves odd-pinnate, leaflets 7～15, blades densely glandular above, densely villous beneath; racemes axillary, longer than leaves, with 10～20 flowers; papilionaceous corolla red or reddish-purple; pods with 1～4 loments, slightly inflated, densely pubescent and prickly. Flowering June to July.
Distribution: Inner Mongolia (Helan Mountains), Shaanxi, Ningxia and Gansu
Habitat: Ravines in desert areas
Use: Forage; ornamental

驼蹄瓣 (豆型霸王)
Zygophyllum fabago L.
蒺藜科
Zygophyllaceae

【特征】多年生肉质草本，高20～80cm；茎多分枝，枝条开展或铺散；小叶1对，倒卵形或矩圆状倒卵形，长15～33mm，宽6～20mm；花1～2枚腋生；萼片绿色，边缘白色膜质；花瓣5，上部近白色，下部橘红色；蒴果矩圆状或圆柱状，具棱，无翅。花期5～6月。

【分布】内蒙古西部、甘肃西北部、青海、新疆；阿富汗、巴基斯坦、中亚、西亚、北非、欧洲东南部。

【生境】荒漠地带的洪积平原、绿洲、河谷、湿润沙地和荒地。

【用途】饲用。

Syrian beancaper
Caltrop family
Perennial fleshy herb 20～80 cm tall; stems much branched, branches spreading or diffuse; leaves with leaflets in 1 pairs, blades obovate or oblong-obovate, 15～33 mm long, 6～20 mm wide; 1 or 2 flowers axillary; sepals green, margins white-membranous; petals 5, apically whitish, basally orange-red; capsules oblong or cylindric, ribbed, wingless. Flowering May to June.

Distribution: W Inner Mongolia, NW Gansu, Qinghai and Xinjiang; Afghanistan, Pakistan, Central Asia, W Asia, N Africa, SE Europe

Habitat: Alluvial plains, oasis, valleys, moist sands and wastelands in desert zone

Use: Forage

戈壁驼蹄瓣 (戈壁霸王)
Zygophyllum gobicum Maxim.
蒺藜科
Zygophyllaceae

【特征】多年生肉质草本；茎有时带橘红色，由基部多分枝，枝平卧，长10～20cm；小叶1对，倒卵形、斜倒卵形或斜卵形；花2枚并生于叶腋；萼片绿色至橘红色；花瓣5，短于萼片，淡绿色或橘红色；蒴果两端钝圆，长为宽的2倍以下，无翅。花期6月。

【分布】内蒙古西部、甘肃西北部、新疆东部；哈萨克斯坦、蒙古。

【生境】戈壁荒漠。

【用途】饲用。

Gobi beancaper
Caltrop family
Perennial fleshy herb; stems sometimes orangish, much branched from base, branches prostrate, 10～20 cm long; leaves with leaflets in 1 pairs, blades obovate, obliquely obovate or obliquely ovate; paired flowers axillary; sepals green to orange-red; petals 5, shorter than sepals, pale green or orange-red; capsules obtuse both ends, the length less than twice the width, wingless. Flowering June.
Distribution: W Inner Mongolia, NW Gansu and E Xinjiang; Kazakhstan, Mongolia
Habitat: Gobi desert
Use: Forage

大花驼蹄瓣 (大花霸王)
Zygophyllum potaninii Maxim.
蒺藜科
Zygophyllaceae

【特征】多年生肉质草本，高10～25cm；茎直立或开展，由基部多分枝；小叶1～2对，斜倒卵形或近圆形；花2～3枚腋生；萼片绿色或带橘黄色；花瓣5，短于萼片，上部白色，下部橘红色；蒴果卵状球形或近球形，具5宽翅，翅宽约6mm。花期5～6月。
【分布】内蒙古西部、甘肃西北部、新疆北部和东部；蒙古、中亚。
【生境】荒漠地带的石质和砾质坡地、残丘、戈壁。
【用途】饲用。

Bigflower beancaper
Caltrop family
Perennial fleshy herb, 10～25 cm tall; stems erect or spreading, much branched from base; leaves with leaflets in 1 or 2 pairs, blades obliquely obovate or suborbicular; 2 or 3 flowers axillary; sepals green or orangish; petals 5, shorter than sepals, apically white, basally orange-red; capsules ovoid-globose or subglobose, with 5 broad wings to 6 mm wide. Flowering May to June.
Distribution: W Inner Mongolia, NW Gansu, N and E Xinjiang; Mongolia, Central Asia
Habitat: Stony and gravelly hillsides, monadnocks and gobi in desert zone
Use: Forage

上图拍摄人：赵利清 Upper photo by Zhao Liqing

翼果驼蹄瓣（翼果霸王）
Zygophyllum pterocarpum Bunge
蒺藜科
Zygophyllaceae

【特征】多年生肉质草本，高10～20cm；茎多数，细弱，开展；小叶2～3对，条状矩圆形或倒披针形；花1～2枚腋生；萼片绿色或带橘黄色；花瓣5，稍长于萼片，上部白色，下部橘红色；蒴果矩圆状卵形或卵形，具5狭翅，翅宽2～3mm。花期6～7月。

【分布】内蒙古西部、甘肃、新疆；蒙古、中亚。

【生境】荒漠地带的石质和砾质坡地、干河床、戈壁。

【用途】饲用。

Wingfruit beancaper
Caltrop family
Perennial fleshy herb, 10～20 cm tall; stems many, slender, spreading; leaves with leaflets in 2 or 3 pairs, blades linear-oblong or oblanceolate; 1 or 2 flowers axillary; sepals green or orangish; petals 5, slightly longer than sepals, apically white, basally orange-red; capsules oblong-ovoid or ovoid, with 5 narrow wings to 2～3 mm wide. Flowering June to July.
Distribution: W Inner Mongolia, Gansu and Xinjiang; Mongolia, Central Asia
Habitat: Stony and gravelly hillsides, dry riverbeds and gobi in desert zone
Use: Forage

石生驼蹄瓣 (石生霸王)
Zygophyllum rosowii Bunge
蒺藜科
Zygophyllaceae

【特征】多年生肉质草本，高10～20cm；茎由基部多分枝，通常开展；小叶1对，卵形、近圆形或矩圆形；花1～2枚腋生；萼片带橘黄色，边缘膜质；花瓣5，与萼片近等长，上部白色，下部橘红色；蒴果先端渐尖，长为宽的2倍以上，无翅。花期5～6月。
【分布】内蒙古西部、甘肃西北部、新疆；蒙古、中亚。
【生境】荒漠地带的砾石山坡、峭壁、碎石地和沙质地。
【用途】饲用。

Rock beancaper
Caltrop family
Perennial fleshy herb, 10～20 cm tall; stems much branched from base, usually spreading; leaves with leaflets in 1 pairs, blades ovate, suborbicular or oblong; 1 or 2 flowers axillary; sepals orangish, margins membranous; petals 5, subequaling sepals, apically white, basally orange-red; capsules acuminate at apex, the length over twice the width, wingless. Flowering May to June.
Distribution: W Inner Mongolia, NW Gansu and Xinjiang; Mongolia, Central Asia
Habitat: Gravelly slopes, cliffs, broken-stony and sandy grounds in desert zone
Use: Forage

毛百合
Lilium dauricum Ker Gawl.
百合科
Liliaceae

【特征】多年生草本,高30～120cm;具鳞茎;茎直立;叶条形或条状披针形,基部有1簇白绵毛;花1～3枚顶生;花被片6,直立,红色,有紫红色斑点;花柱长于子房2倍以上。花期6～7月。
【分布】东北,河北,内蒙古东部;俄罗斯(西伯利亚、远东)。
【生境】山地灌丛、疏林、草甸。
【用途】饲用;观赏。

Candlestick lily
Lily family
Perennial herb 30～120 cm tall; bulbiferous; stems erect; leaves linear or linear-lanceolate, base white-lanate in a cluster; flowers 1～3, terminal; tepals 6, erect, red with purple-red spots; style over twice longer than the ovary. Flowering June to July.
Distribution: NE China, Hebei and E Inner Mongolia; Russia (Siberia and Far East)
Habitat: Montane scrublands, open woodlands and meadows
Use: Forage; ornamental

拍摄人:易津 Photo by Yi Jin

石竹
Dianthus chinensis L.
石竹科
Caryophyllaceae

【特征】多年生草本，高20～40cm；茎疏丛生，直立，光滑无毛；叶对生；花单生或2～3枚组成聚伞花序；苞片2～3对，卵形，先端尾尖，长为萼的1/2；花瓣5，紫红色、粉色或白色，上端边缘有齿，喉部有深色圈纹和须毛。花期6～8月。

【分布】东北、华北、西北及长江流域；朝鲜、俄罗斯（远东）。

【生境】山地草甸、草甸草原。

【用途】饲用；观赏；药用。

Rainbow pink
Pink family
Perennial herb 20～40 cm tall; stems loosely caespitose, erect, glabrous; leaves opposite; flowers solitary or 2～3 in a cyme; bracts in 2～3 pairs, ovate, apex caudate, 1/2 as long as the calyx; petals 5, purple-red, pink or white, toothed at upper margin, dark purple-ringed and bearded at throat. Flowering June to August.

Distribution: NE, N, NW China, and regions along Changjiang River; Korea, Russia (Far East)

Habitat: Montane meadows and meadow-steppe

Use: Forage; ornamental; medicine

簇茎石竹
Dianthus repens Willd.
石竹科
Caryophyllaceae

【特征】多年生草本，高达30cm；茎密丛生，直立或上升，光滑无毛；叶对生；花1或2枚；苞片1～2对，外面1对条形，叶状，长于或近等长于萼，内面1对卵状披针形，比萼短；花瓣5，紫红色，上端边缘有齿，喉部有深色圈纹和须毛。花期6～8月。

【分布】内蒙古；俄罗斯（西伯利亚、远东）、北美洲。

【生境】山地草甸、草甸草原。

【用途】饲用；观赏。

Boreal carnation
Pink family
Perennial herb to 30 cm tall; stems strongly caespitose, erect or ascending, glabrous; leaves opposite; flowers 1 or 2; bracts in 1～2 pairs, outer pairs linear, leaf-like, longer than or subequaling calyx, inner pairs ovate-lanceolate, shorter than calyx; petals 5, purple-red, toothed at upper margin, dark purple-ringed and bearded at throat. Flowering June to August.
Distribution: Inner Mongolia; Russia (Siberia and Far East), North America
Habitat: Montane meadows and meadow-steppe
Use: Forage; ornamental

瞿麦
Dianthus superbus L.
石竹科
Caryophyllaceae

【特征】多年生草本，高30～50cm；茎丛生，直立，上部稍分枝；叶对生；聚伞花序顶生，或花单生；花瓣5，淡紫红色，稀白色，上端细裂成流苏状，喉部具须毛。花期7～8月。

【分布】东北、华北、西北、华东，四川；蒙古、朝鲜、日本、俄罗斯、欧洲。

【生境】林下、林缘、草甸、沟谷、溪边。

【用途】饲用；观赏；药用。

Fringed pink
Pink family
Perennial herb, 30～50 cm tall; stems caespitose, erect, slightly branched above; leaves opposite; flowers in a terminal cyme, or solitary; petals 5, pale purplish-red, rarely white, fimbriate above, bearded at throat. Flowering July to August.
Distribution: NE, N, NW and E China, Sichuan; Mongolia, Korea, Japan, Russia, Europe
Habitat: Forests, forest edges, meadows, gullies and streamsides
Use: Forage; ornamental; medicine

拍摄人：赵 凡 Photos by Zhao Fan

拍摄人：张洪江 Photos by Zhang Hongjiang

新疆芍药
Paeonia sinjiangensis K. Y. Pan
芍药科 (牡丹科)
Paeoniaceae

【特征】多年生草本，高40～80cm；根分枝圆柱形；叶为1至2回三出复叶，小叶羽状分裂，裂片披针形至狭披针形，全缘；花单生茎顶；萼片5；花瓣9，紫红色；心皮4～5，稀2或3；蓇葖果被毛，成熟时毛脱落。花期6～7月。
【分布】新疆北部。
【生境】山地林下、山坡草地。
【用途】观赏；药用。

Xinjiang peony
Peony family
Perennial herb 40～80 cm tall; root branches cylindric; leaves ternate to biternate, leaflets pinnatipartite, segments lanceolate to narrowly lanceolate, entire; solitary flower terminal; sepals 5; petals 9, purplish-red; carpels 4 to 5, rarely 2 or 3; follicles hairy, glabrate at maturity. Flowering June to July.
Distribution: N Xinjiang
Habitat: Montane forests and sloping grasslands
Use: Ornamental; medicine

刺蔷薇 (大叶蔷薇)
Rosa acicularis Lindl.
蔷薇科
Rosaceae

【特征】灌木，高约1m；多分枝，枝红褐色，密生皮刺；单数羽状复叶，小叶5~7，小叶片长2~5cm，边缘有锯齿；花单生叶腋；萼片外面被柔毛和腺毛；花瓣5，玫瑰红色；蔷薇果椭圆状至梨形，红色，光滑，宿存萼片直立。花期6~7月。

【分布】东北、华北、西北；蒙古、朝鲜、日本、俄罗斯、北欧、北美洲。

【生境】林下、林缘、山地灌丛。

【用途】观赏；药用。

Prickly rose
Rose family
Shrub to 1 m tall; much branched, branches red-brown, densely prickly; leaves odd-pinnate, leaflets 5~7, leaflet blades 2~5 cm long, serrate; solitary flower axillary; sepals pubescent and glandular-hairy outer side; petals 5, rosy-red; hips ellipsoid to pyriform, red, smooth, persistent sepals erect. Flowering June to July.
Distribution: NE, N and NW China; Mongolia, Korea, Japan, Russia, N Europe, North America
Habitat: Forests, forest margins and montane scrublands
Use: Ornamental; medicine

拍摄人：易 津（右图）&赵利清（左上图） Photos by Yi Jin & Zhao Liqing

拍摄人：赵利清 Photos by Zhao Liqing

美蔷薇
Rosa bella Rehder et E. H. Wilson
蔷薇科
Rosaceae

【特征】灌木，高1~3m；枝常带紫色，疏生皮刺；单数羽状复叶，小叶7~9，稀5，边缘有锯齿；花单生或2~3枚簇生；花梗和花萼密被腺毛；花瓣5，粉红色或紫红色；蔷薇果椭圆状或矩圆状，鲜红色，密被腺状刚毛，宿存萼片直立。花期6~7月。
【分布】华北，吉林、河南。
【生境】山地林缘、沟谷、陡崖。
【用途】观赏；水土保持；提取芳香油；药用。

Solitary rose
Rose family
Shrub 1~3 m tall; branches usually purplish, sparsely prickly; leaves odd-pinnate, leaflets 7~9, rarely 5, serrate; flowers solitary or in clusters of 2~3; pedicel and calyx densely glandular-hairy; petals 5, pink or purple-red; hips ellipsoid or oblong, bright-red, with dense glandular-bristles, persistent sepals erect. Flowering June to July.
Distribution: N China, Jilin and Henan
Habitat: Montane forest margins, ravines and steep slopes
Use: Ornamental; soil conservation; extracting essential oil; medicine

荒漠黄耆
Astragalus alaschanensis H. C. Fu
(*Astrasgalus dengkouensis* H. C. Fu)

豆科
Fabaceae (Leguminosae)

【特征】多年生草本，高10～20cm，全株密被丁字毛；茎极短缩；单数羽状复叶，小叶11～25，小叶片宽椭圆形至近圆形；总状花序密集于叶丛基部；蝶形花冠紫红色，龙骨瓣略短于翼瓣；荚果卵状或矩圆状卵形，稍膨胀，密被长硬毛。花期5～6月。
【分布】内蒙古西部、宁夏、甘肃。
【生境】荒漠和荒漠草原地带的沙地。
【用途】饲用。

Desert milkvetch
Pea family
Perennial herb 10～20 cm tall, densely medifixed-hairy throughout; stems strongly shortened; leaves odd-pinnate, leaflets 11～25, blades broadly elliptic to suborbicular; racemes congested in base of leaf cluster; papilionaceous corolla purple-red, keel slightly shorter than wings; pods ovoid or oblong-ovoid, slightly inflated, densely hirsute. Flowering May to June.
Distribution: W Inner Mongolia, Ningxia and Gansu
Habitat: Sands in desert and desert-steppe zones
Use: Forage

粗壮黄耆(乌拉特黄耆、贺兰山黄耆)
Astragalus hoantchy Franch.
豆科
Fabaceae (Leguminosae)

【特征】多年生草本，高达1m；茎直立，多分枝；单数羽状复叶，小叶9～25，小叶片宽卵形至近圆形；总状花序腋生，长20～30cm，具花12～15枚；蝶形花冠紫红色或紫色；荚果矩圆状，下垂，具长柄。花期5～6月。
【分布】内蒙古西部、宁夏、甘肃、青海东部。
【生境】海拔1 400～2 400m的山坡、沟谷、水边。
【用途】饲用。

Wulate milkvetch
Pea family
Perennial herb up to 1 m tall; stems erect, much branched; leaves odd-pinnate, leaflets 9～25, blades broadly elliptic to suborbicular; racemes axillary, 20～30 cm long, with 12～15 flowers; papilionaceous corolla purple-red or purple; pods oblong, nodding, with a long stipe. Flowering May to June.
Distribution: W Inner Mongolia, Ningxia, Gansu and E Qinghai
Habitat: Slopes, ravines and watersides at 1 400～2 400 m
Use: Forage

了墩黄耆
Astragalus pavlovii B. Fedtsch. et Basil.
(*Astragalus lioui* Tsai et Yu)
豆科
Fabaceae (Leguminosae)

【特征】一年生草本，高10～25cm；茎直立或稍斜升，有分枝；单数羽状复叶，小叶5～7，小叶片矩圆状倒卵形或矩圆状倒披针形；短总状花序腋生，花5～25枚；蝶形花冠紫红色或淡紫色，翼瓣先端微凹；荚果矩圆状，光滑，背面有窄沟。花期5～7月。
【分布】内蒙古西部、宁夏、甘肃、新疆；蒙古。
【生境】荒漠区的干河床、洼地和沙砾质地。
【用途】饲用。

Liu milkvetch
Pea family
Annual herb 10～25 cm tall; stems erect or slightly ascending, branched; leaves odd-pinnate, leaflets 5～7, blades oblong-obovate or oblong-oblanceolate; short racemes axillary, with 5～25 flowers; papilionaceous corolla purple-red or pale purple, wings retuse at apex; pods oblong, smooth, back narrowly fluted. Flowering May to July.
Distribution: W Inner Mongolia, Ningxia, Gansu and Xinjiang; Mongolia
Habitat: Dry riverbeds, swales, sandy and gravelly sites in desert areas
Use: Forage

甘肃旱雀豆
Chesniella ferganensis (Korsh.) Boriss.
(戈壁雀儿豆 *Chesneya grubovii* Yakovl.)
豆科
Fabaceae (Leguminosae)

【特征】多年生草本，全株密被柔毛；茎平卧，长2~20cm；单数羽状复叶，小叶7~11，小叶片倒卵形，先端截形，具刺尖，两面密被柔毛；花单生于叶腋；蝶形花冠粉红色或紫红色，旗瓣背面密被柔毛；荚果矩圆状，密被长柔毛。花期6~7月。
【分布】内蒙古西部、甘肃西部；蒙古。
【生境】荒漠地带的砾石质山坡和戈壁。
【用途】饲用。

Gansu drybirdbean
Pea family
Perennial herb, densely pubescent throughout; stems prostrate, 2~20 cm long; leaves odd-pinnate, leaflets 7~11, blades obovate, apex truncate, spine-pointed, densely pubescent; flowers solitary in leaf axils; papilionaceous corolla pink or purple-red, banner densely pubescent on back; pods oblong, densely villous. Flowering June to July.
Distribution: W Inner Mongolia and W Gansu; Mongolia
Habitat: Gravelly slopes and gobi in desert zone
Use: Forage

盐豆木 (铃铛刺)
Halimodendron halodendron (Pall.) Voss
豆科
Fabaceae (Leguminosae)

【特征】灌木，高50~200cm；多分枝，当年生小枝密被柔毛；羽状复叶，托叶和叶轴针刺状，小叶2~6，倒披针形；总状花序具花2~3枚；蝶形花冠淡紫色至紫红色；荚果矩圆状倒卵形，革质，膨胀。花期5~7月。
【分布】内蒙古西部、甘肃、新疆；蒙古、西伯利亚、中亚、高加索。
【生境】荒漠区盐化沙地、河岸。
【用途】饲用；固沙；绿篱。

Common salttree
Pea family
Shrub 50~200 cm tall; much branched, new branchlets densely pubescent; leaves pinnate, stipules and rachis spine-like, leaflets 2~6, oblanceolate; racemes with 2~3 flowers; papilionaceous corolla pale purple to purple-red; pods oblong-obovoid, leathery, inflated. Flowering May to July.
Distribution: W Inner Mongolia, Gansu and Xinjiang; Mongolia, Siberia, Central Asia, Caucasia
Habitat: Saline sands and river banks in desert areas
Use: Forage; fixing dunes; green fences

拍摄人：赵利清 Photos by Zhao Liqing

线棘豆
Oxytropis filiformis DC.
豆科
Fabaceae (Leguminosae)

【特征】多年生草本，高10～20cm；茎无或极短缩；单数羽状复叶，小叶17～31（45），小叶片条状披针形至卵状披针形，长约5mm，宽1～2mm，两面被柔毛；总状花序长2.5～5cm，花10～15枚；蝶形花冠蓝紫色或紫红色；荚果宽椭圆状至卵状，硬膜质，疏被短毛。花期6～7月。
【分布】东北，内蒙古；蒙古、西伯利亚。
【生境】石质山坡、丘陵、草甸。
【用途】饲用。

Slender locoweed
Pea family
Perennial herb 10～20 cm tall; acaulescent or stems strongly shortened; leaves odd-pinnate, leaflets 17～31 (45), blades linear-lanceolate to ovate-lanceolate, about 5 mm long, 1～2 mm wide, pubescent; racemes 2.5～5 cm long, with 10～15 flowers; papilionaceous corolla bluish-purple or purple-red; pods broadly ellipsoid to ovoid, hard-membranous, sparsely short-hairy. Flowering June to July.
Distribution: NE China, Inner Mongolia; Mongolia, Siberia
Habitat: Rocky slopes, hills and meadows
Use: Forage

红花海绵豆
Spongiocarpella grubovii (Ulzij.) Yakcvl.
(大花雀儿豆 *Chesneya macrantha* S. H. Cheng ex H. C. Fu)
豆科
Fabaceae (Leguminosae)

【特征】垫状半灌木，高5～15cm；茎短缩，多分枝；单数羽状复叶，叶轴针刺状，小叶7～11，小叶片两面被白色绢毛，上面被褐色腺点；花单生于叶腋；花萼密被长柔毛和褐色腺体，萼筒基部囊状；蝶形花冠紫红色；荚果圆柱状，果皮海绵质。花期5～6月。
【分布】内蒙古西部；蒙古。
【生境】荒漠及荒漠草原带的山地石缝、剥蚀残丘、沙地。
【用途】饲用；观赏。

Bigflower birdbean
Pea family
Cushion-like subshrub 5～15 cm tall; stems shortened, densely branched; leaves odd-pinnate, rachis spine-like, leaflets 7～11, blades white-sericous both sides, brown-glandular above; flowers solitary in leaf axils; calyx densely villous and brown-glandular, calyx tube basally saccate; papilionaceous corolla purple-red; pods cylindric, pericarp spongiose. Flowering May to June.
Distribution: W Inner Mongolia; Mongolia
Habitat: Mountain rock crevices, eroded hills and sands in desert and desert-steppe zones
Use: Forage; ornamental

河北假报春
Cortusa matthioli ssp. *pekinensis* (V. A. Richt.) Kitag.
报春花科
Primulaceae

【特征】多年生草本；叶基生，具长柄，叶柄被长柔毛，叶片轮廓近圆形或圆肾形，掌状7~11浅裂至半裂，边缘具不整齐疏齿；花葶高20~30cm；伞形花序具花6~11枚；花冠紫红色，漏斗状钟形，裂片5。花期6~7月。

【分布】华北；朝鲜、俄罗斯（远东）。

【生境】山地林下、林缘、灌丛、溪边。

【用途】饲用；观赏。

Hebei alpine bells
Primula family
Perennial herb; leaves basal, petioles long and villous, blades suborbicular or rounded-reniform in outline, palmately 7- to11-lobed to cleft, irregularly spaced-serrate; scapes 20~30 cm tall; umbels with flowers 6~11; corolla purple-red, infundibular-campanulate, with 5 lobes. Flowering June to July.
Distribution: N China; Korea, Russia (Far East)
Habitat: Montane forests, forest margins, thickets and streamsides
Use: Forage; ornamental

拍摄人：赵利清 Photos by Zhao Liqing

翠南报春 (樱草)
Primula sieboldii E. Morren
报春花科
Primulaceae

【特征】多年生草本；基生叶莲座状，具柄，叶片卵形至矩圆形，边缘浅裂，裂片具圆齿，两面被长柔毛；花葶高12～30cm；伞形花序顶生，具花2～15枚；花冠紫红色或淡紫色，稀白色，高脚碟状，裂片5，裂片先端2深裂。花期5～6月。

【分布】东北、内蒙古；朝鲜、日本、俄罗斯（西伯利亚、远东）。

【生境】草甸、沼泽、山地林下及林缘。

【用途】饲用；观赏。

Siebold primrose
Primula family
Perennial herb; basal leaves rosulate, petiolate, blades ovate to oblong, margins lobed, lobes crenate, villous both sides; scapes 12～30 cm tall; umbels terminal, with 2～15 flowers; corolla purple-red or pale purple, rarely white, salverform, with 5 lobes, lobes 2-parted at apex. Flowering May to June.
Distribution: NE China, Inner Mongolia; Korea, Japan, Russia (Siberia and Far East)
Habitat: Meadows, marshes, montane forests and forest margins
Use: Forage; ornamental

拍摄人：赵利清 Photos by Zhao Liqing

下图拍摄人：赵利清 Lower photo by Zhao Liqing

杠柳
Periploca sepium Bunge
萝藦科
Asclepiadcaeae

【特征】藤状灌木，高约1m，具乳汁；茎散生；叶对生，革质；二歧聚伞花序，花数枚；花冠辐状，瓣片5，紫红色，中央加厚呈纺锤形，反折，内被长柔毛，外面无毛；副花冠具5被毛的弯钩；蓇葖果2，圆筒状，弯曲，长8～11cm，顶端相连，略有光泽。花期6～7月。

【分布】辽宁、内蒙古、河北、山东、河南、陕西、宁夏、甘肃、四川、贵州；俄罗斯（远东）。

【生境】黄土丘陵、沙丘或沙地。

【用途】纤维材料；药用；制杀虫剂。

Chinese silkvine
Milkweed family
Vine-like shrub to 1 m tall, with milky juice; stems diffuse; leaves opposite, leathery; dichasium with several flowers; corolla rotate, segments 5, purple-red, center thickened to fusiform, reflexed, villous inside, glabrous outside; corona with 5 hairy hooks; follicles 2, cylindric, curved, 8～11 cm long, apically linked, slightly shiny. Flowering June to July.
Distribution: Liaoning, Inner Mongolia, Hebei, Shandong, Henan, Shaanxi, Ningxia, Gansu, Sichuan and Guizhou; Russia (Far East)
Habitat: Loess hills, dunes or sands
Use: Fiber materials; medicine; making insecticide

返顾马先蒿
Pedicularis resupinata L.
玄参科
Scrophulariaceae

拍摄人：赵 凡 Photos by Zhao Fan

【特征】多年生草本，高30～70cm；茎直立，上部分枝，具四棱，带紫色；叶披针形至狭卵形，边缘具钝圆重齿，常反卷；总状花序顶生；花萼近无毛；花冠紫红色，二唇形，花管扭曲，上唇盔状，先端伸长为喙，下唇稍长于上唇，3浅裂。花期6～8月。

【分布】东北、华北、华东，陕西、甘肃、四川；蒙古、朝鲜、日本、俄罗斯（西伯利亚、远东）、欧洲。

【生境】草甸、林缘。

【用途】药用；观赏。

Upside-down lousewort
Figwort family
Perennial herb 30～70 cm tall; stems erect, branched above, quadrangular, purplish; leaves lanceolate to narrowly ovate, margins duplicato-crenate, usually revolute; racemes terminal; calyx subglabrate; corolla purplish-red, bilabiate, tube twisted, upper lip galeate, apex protruding into a beak, lower lip slightly longer than the upper lip, 3-lobed. Flowering June to August.
Distribution: NE, N and E China, Shaanxi, Gansu and Sichuan; Mongolia, Korea, Japan, Russia (Siberia and Far East), Europe
Habitat: Meadows and forests edges
Use: Medicine; ornamental

毛返顾马先蒿
Pedicularis resupinata var. *pubescens* Nakai
玄参科
Scrophulariaceae

【特征】与返顾马先蒿(*Pedicularis resupinata*)的区别为：茎、叶，苞片、花萼密被白色柔毛。
【分布】东北，内蒙古东部；朝鲜、日本。
【生境】山地和河谷，林间、林缘和草甸。
【用途】观赏。

Hairy upside-down lousewort
Figwort family
Difference to *Pedicularis resupinata*: Stems, leaves, bracts and calyx densely white-pubescent.
Distribution: NE China, E Inner Mongolia; Korea, Japan
Habitat: Mountains and valleys, forests, forest margins and meadows
Use: Ornamental

拍摄人：赵利清　Photo by Zhao Liqing

华北马先蒿 (塔氏马先蒿)
Pedicularis tatarinowii Maxim.
玄参科
Scrophulariaceae

【特征】一年生草本，高8~40cm；茎直立或斜升，圆柱形，具4条纵毛线，分枝轮生；叶轮生，叶片羽状全裂，裂片羽状浅裂或深裂；总状花序顶生；花萼膨大，膜质，被毛；花冠紫红色，二唇形，上唇盔状，顶部弓曲，喙强烈弯曲，下唇长于上唇，3浅裂。花期7~9月。
【分布】华北特有种。
【生境】山地草甸和林缘草甸。
【用途】观赏。

North China lousewort
Figwort family
Annual herb 8~40 cm tall; stems erect or ascending, cylindric, with 4 lines of hairs, branches whorled; leaves whorled, blades pinnatisect, segments pinnately lobed or parted; racemes terminal; calyx inflated, membranous, hairy; corolla purplish-red, bilabiate, upper lip galeate, apex bent, with a beak strongly bent, lower lip longer than the upper lip, 3-lobed. Flowering July to September.
Distribution: Endemic to N China
Habitat: Montane meadows, and meadows along forest edges
Use: Ornamental

拍摄人：赵利清 Photo by Zhao Liqing

拍摄人：赵 凡 Photos by Zhao Fan

轮叶马先蒿
Pedicularis verticillata L.
玄参科
Scrophulariaceae

【特征】多年生草本，高15～35cm；茎直立，具4条纵毛线；茎生叶4枚轮生，叶片羽状深裂至全裂，裂片具缺刻状齿；总状花序顶生；花萼膨大，膜质，密被长柔毛；花冠紫红色，二唇形，上唇盔状，略弓曲，下唇稍长或等长于上唇，3浅裂。花期6～7月。

【分布】东北、华北、西北、四川、西藏；北半球寒温带及北极。

【生境】草甸、沼泽草甸、林间灌丛、冻原。

【用途】观赏。

Whorled lousewort
Figwort family
Perennial herb 15～35 cm tall; stems erect, with 4 lines of hairs; cauline leaves in whorls of 4, blades pinnately parted to divided, segments incised-toothed; racemes terminal; calyx inflated, membranous, densely villous; corolla purplish-red, bilabiate, upper lip galeate, slightly bent, lower lip slightly longer than or equaling the upper lip, 3-lobed. Flowering June to July.
Distribution: NE, N and NW China, Sichuan and Tibet; cool temperate zone in the Northern Hemisphere, the Arctic
Habitat: Meadows, swamp-meadows, thickets in forests, and tundra
Use: Ornamental

上图拍摄人：赵利清 Upper photo by Zhao Liqing

内蒙野丁香
Leptodermis ordosica H. C. Fu et E. W. Ma
茜草科
Rubiaceae

【特征】小灌木，高20～40cm；多分枝；叶对生或假轮生，宽椭圆形至狭椭圆形，全缘，边缘常反卷；花1～3枚簇生于叶腋或枝顶；花冠长漏斗状，紫红色，裂片4～5；蒴果椭圆状。花期7～8月。
【分布】内蒙古。
【生境】荒漠中的山坡岩缝。
【用途】庭院绿化。

Ordos wildclove
Madder family
Small shrub 20～40 cm tall; much branched; leaves opposite or pseudo-whorled, widely or narrowly elliptic, entire, margins usually revolute; flowers 1～3 clustered in axils or on branch top; corolla long-funnelform, purple-red, with 4 or 5 lobes; capsules ellipsoid. Flowering July to August.
Distribution: Inner Mongolia
Habitat: Sloping rock crevices in desert
Use: Courtyard planting

牛蒡
Arctium lappa L.
菊科
Asteraceae (Compositae)

【特征】二年生草本，高1～2m；茎直立，粗壮，带紫色，上部多分枝；基生叶大型，叶片宽卵形或心形，长达30cm，全缘或边缘浅波状，下面密被绵毛，上部叶渐小；头状花序排列成伞房状；总苞片顶端有倒钩刺；管状小花紫红色。花期6～8月。
【分布】全国各地；欧亚大陆广布。
【生境】山坡、山谷、林间、林缘、荒地、河边、路旁、村庄附近，海拔750～3 500m。
【用途】果实和根药用。

Great burdock
Aster family
Biennial herb 1～2 m tall; stems erect, stout, purplish, much branched above; basal leaves large, blades broadly ovate or cordate, to 30 cm long, entire or repand, densely lanate below, upper leaves reduced; heads corymbiform-arranged; involucral bracts with uncinate spines at apex; tubular florets purple-red. Flowering June to August.
Distribution: Throughout China; widespread in Eurasia
Habitat: Slopes, valleys, forest openings and borders, wastelands, riversides, roadsides and village sides at 750～3 500 m
Use: Fruits and roots for medicine

拍摄人：卢欣石 Photos by Lu Xinshi

丝路蓟
Cirsium arvense (L.) Scop.
菊科
Asteraceae (Compositae)

【特征】多年生草本，高50～160cm；茎直立，被蛛丝状毛；叶羽状浅裂或半裂，裂片边缘有刺齿和针刺，叶下面被蛛丝状绵毛；头状花序排列成圆锥状伞房花序；外层总苞片顶端有针刺；管状小花粉紫色。花期6～7月。
【分布】西北、内蒙古西部；蒙古、中亚、阿富汗、印度、欧洲。
【生境】戈壁、沙地、砾石山坡、河滩、水边、田间、路旁。
【用途】药用；根食用。

Canada thistle (Creeping thistle, Field thistle, Perennial thistle, Prickly thistle)
Aster family
Perennial herb 50～160 cm tall; stems erect, arachnoid-hairy; leaves pinnately lobed or cleft, lobes with spinose teeth and acicular spines at margins, arachnoid-tomentose below; heads in a paniculate corymb; outer involucral bracts acicular-spinose at apex; tubular florets pinkish-purple. Flowering June to July.
Distribution: NW China, W Inner Mongolia; Mongolia, Central Asia, Afghanistan, India, Europe
Habitat: Gobi, sands, gravelly slopes, flood lands, watersides, farmlands and roadsides
Use: Medicine; roots edible

密花风毛菊 (渐尖风毛菊)
Saussurea acuminata Turcz.
菊科
Asteraceae (Compositae)

【特征】多年生草本，高30～60cm；茎直立，单一，不分枝，有狭翅；叶质厚，茎生叶披针形至条状披针形，全缘，基部半抱茎，边缘被糙硬毛，反卷；头状花序多数，密集成半球形伞房状；总苞片先端长尾尖，反折，被柔毛，内层者带紫色；管状小花淡紫色。花期8月。
【分布】东北，内蒙古；蒙古、西伯利亚。
【生境】森林至草原带的湿润草甸。
【用途】饲用。

Tapered saw-wort
Aster family
Perennial herb 30～60 cm tall; stems erect, single, unbranched, narrowly winged; leaves thick, cauline leaves lanceolate to linear-lanceolate, entire, base clasping, margins hispidulous, revolute; many heads in a hemispheric-corymbiform cluster; involucral bracts apically long-caudate, reflexed, pubescent, inner ones purplish; tubular florets pale purple. Flowering August.
Distribution: NE China, Inner Mongolia; Mongolia, Siberia
Habitat: Moist meadows in forest to steppe zones
Use: Forage

拍摄人：赵利清 Photos by Zhao Liqing

荒漠风毛菊
Saussurea deserticola H. C. Fu
菊科
Asteraceae (Compositae)

【特征】多年生草本，高30~40cm；茎单一，直立；叶2回羽状全裂，裂片11~13对，两面被蛛丝状毛和腺点；头状花序多数，圆锥状排列；总苞片革质，黑绿色或褐色，密被蛛丝状柔毛，上部及边缘带紫红色；管状小花粉紫色。花期8~9月。
【分布】内蒙古（千里山）。
【生境】荒漠草原区石质山坡。
【用途】饲用。

Desert saw-wort
Aster family
Perennial herb 30~40 cm tall; stems single, erect; leaves bipinnatisect, segments in 11~13 pairs, arachnoid-pubescent and glandular; many heads paniculately arranged; involucral bracts leathery, black-green or brown, densely arachnoid-pubescent, apically and marginally purplish-red; tubular florets pink-purple. Flowering August to September.
Distribution: Inner Mongolia (Qianli Mountains)
Habitat: Rocky slopes in desert-steppe areas
Use: Forage

硬叶风毛菊 (硬叶乌苏里风毛菊)
Saussurea firma (Kitag.) Kitam.

菊科
Asteraceae (Compositae)

【特征】多年生草本，高50～80cm；茎直立，不分枝；叶革质，基生叶和茎下部叶卵形至宽卵形，边缘有波状齿和短硬毛，上部叶矩圆状卵形至条状，边缘有细锯齿；头状花序多数，排列成伞房状；总苞片不反折，顶端和边缘带紫色，被蛛丝状毛；管状小花紫红色。花期7～8月。
【分布】东北，内蒙古、河北。
【生境】草甸、灌丛。
【用途】饲用。

Stiffleaf saw-wort
Aster family
Perennial herb 50～80 cm tall; stems erect, unbranched; leaves leathery, basal and lower cauline leaves ovate to broadly ovate, margins repand-serrate and hispidulous, upper leaves oblong-ovate to linear, serrulate; many heads corymbose-arranged; involucral bracts not reflexed, apically and marginally purplish, arachnoid-hairy; tubular florets purple-red. Flowering July to August.
Distribution: NE China, Inner Mongolia and Hebei
Habitat: Meadows and thickets
Use: Forage

左图拍摄人：赵利清 Left photo by Zhao Liqing

西北风毛菊
Saussurea petrovii Lipsch.
菊科
Asteraceae (Compositae)

【特征】多年生草本，高15～25cm；茎丛生，直立，密被柔毛；叶条形，边缘有疏齿，上部叶常全缘，叶下面被白色毡毛；头状花序少数，排列成复伞房状；总苞片被蛛丝状短柔毛，边缘带紫色；管状小花粉红色。花期8～9月。
【分布】内蒙古西部、甘肃。
【生境】荒漠草原砾石质地。
【用途】饲用。

Petrov saw-wort
Aster family
Perennial herb 15～25 cm tall; stems tufted, erect, densely pubescent; leaves linear, spaced-serrate, upper leaves usually entire, white-manicate below; few heads compound corymbose-arranged; involucral bracts arachnoid-pubescent, with purplish edges; tubular florets pink. Flowering August to September.
Distribution: W Inner Mongolia and Gansu
Habitat: Gravelly sites in desert-steppe
Use: Forage

折苞风毛菊
Saussurea recurvata (Maxim.) Lipsch.
菊科
Asteraceae (Compositae)

【特征】多年生草本，高40～80cm；茎直立，单生，不分枝；叶质厚，基生叶和茎下部叶卵状三角形至长卵形，不分裂或羽状半裂，边缘具缺刻状锯齿，上部叶披针形或条状披针形，边缘具锯齿或全缘；头状花序数个，密集成伞房状；总苞片反折，带紫色；管状小花紫色。花期7～8月。
【分布】东北、内蒙古、河北、陕西、宁夏、甘肃、青海；朝鲜、俄罗斯（远东）。
【生境】林缘、灌丛、草甸。
【用途】饲用。

Bent saw-wort
Aster family
Perennial herb 40～80 cm tall; stems erect, single, unbranched; leaves thick, basal and lower cauline leaves ovate-triangular to long-ovate, unlobed or pinnatifid, margins incised-serrate, upper leaves lanceolate or linear-lanceolate, serrate or entire; several heads in a corymbiform cluster; involucral bracts reflexed, purplish; tubular florets purple. Flowering July to August.
Distribution: NE China, Inner Mongolia, Hebei, Shaanxi, Ningxia, Gansu and Qinghai; Korea, Russia (Far East)
Habitat: Forest margins, thickets and meadows
Use: Forage

柳叶风毛菊
Saussurea salicifolia (L.) DC.
菊科
Asteraceae (Compositae)

【特征】多年生草本，高15~40cm；茎丛生，直立，被蛛丝状柔毛；叶条形至条状披针形，全缘，稀基部边缘具疏齿，常反卷，下面被白色毡毛；头状花序排列成复伞房状；总苞片紫红色，疏被蛛丝状毛；管状小花粉红色或红紫色。花期8~9月。
【分布】东北，内蒙古；蒙古、西伯利亚。
【生境】草原。
【用途】饲用。

Willowleaf saw-wort
Aster family
Perennial herb 15~40 cm tall; stems tufted, erect, arachnoid-pubescent; leaves linear to linear-lanceolate, entire, rarely spaced-serrate at basal margins, usually revolute, white-manicate below; heads compound corymbose-arranged; involucral bracts purple-red, sparsely arachnoid-hairy; tubular florets pink or red-purple. Flowering August to September.
Distribution: NE China, Inner Mongolia; Mongolia, Siberia
Habitat: Steppe
Use: Forage

拍摄人：赵利清 Photos by Zhao Liqing

雅布赖风毛菊
Saussurea yabulaiensis Y. Y. Yao

菊科
Asteraceae (Compositae)

【特征】多年生草本，高30～35cm；茎直立或弯曲，细长；下部叶羽状全裂，裂片2～3对，条形或披针形，叶柄基部扩大，中、上部叶渐小，不分裂；头状花序少数，排列成伞房状；总苞片革质，边缘疏生腺毛；管状小花粉紫色。花期9～10月。
【分布】内蒙古（阿拉善）。
【生境】荒漠地带的石质山坡和沟谷。
【用途】饲用。

Yabulai saw-wort
Aster family
Perennial herb 30～35 cm tall; stems erect or curved, slender; lower leaves pinnatisect, segments in 2～3 pairs, linear or lanceolate-linear, petiole basally enlarged, middle and upper leaves reduced, undivided; few heads corymbiform-arranged; involucral bracts leathery, margins sparsely glandular-hairy; tubular florets pink-purple. Flowering September to October.
Distribution: Inner Mongolia (Alashan)
Habitat: Rocky slopes and ravines in desert zone
Use: Forage

拍摄人：赵 凡 Photos by Zhao Fan

伪泥胡菜
Serratula coronata L.
菊科
Asteraceae (Compositae)

【特征】多年生草本，高50～100cm；茎直立，上部分枝；叶羽状深裂或全裂，边缘有疏齿和刚毛；头状花序少数，在茎枝顶排列成伞房状，稀单生；总苞钟状，总苞片紫褐色，密被绒毛；管状小花紫红色。花期7～9月。
【分布】东北、华北，山东、江苏、湖北、贵州、新疆；蒙古、日本、俄罗斯（西伯利亚、远东）、中亚、欧洲。
【生境】草甸、林缘。
【用途】饲用；药用。

Crowned false saw-wort
Aster family
Perennial herb 50～100 cm tall; stems erect, branched above; leaves pinnately parted or divided, margins spaced-serrate and bristled; heads few, terminally corymbose-arranged, rarely solitary; involucre campanulate, involucral bracts purple-brown, densely tomentose; tubular florets purple-red. Flowering July to September.
Distribution: NE and N China, Shandong, Jiangsu, Hubei, Guizhou and Xinjiang; Mongolia, Japan, Russia (Siberia and Far East), Central Asia, Europe
Habitat: Meadows and forest edges
Use: Forage; medicine

新疆野百合
Lilium martagon var. *pilosiusculum* Freyn
百合科
Liliaceae
【特征】多年生草本；具鳞茎；茎高30~60cm，光滑；叶轮生，披针形；花单生或数枚排列成总状花序；花冠钟状，下垂；花被片6，反卷，紫红色，有深色斑点。花期5月。
【分布】新疆。
【生境】海拔1 500~2 000m的山地草原及云杉林下。
【用途】观赏；饲用；鳞茎食用。

Xinjiang lily
Lily family
Perennial herb; bulbiferous; stems 30~60 cm tall, glabrous; leaves whorled, lanceolate; flowers solitary or several in a raceme; corolla campanulate, nodding; tepals 6, reflexed, purple-red, dark-mottled. Flowering May.
Distribution: Xinjiang
Habitat: Montane steppe and spruce forests at 1 500~2 000 m
Use: Ornamental; forage; bulbs edible

拍摄人：张洪江 Photos by Zhang Hongjiang

拍摄人：赵利清 Photos by Zhao Liqing

厚叶花旗杆
Dontostemon crassifolius (Bunge) Maxim.
十字花科
Brassicaceae

【特征】多年生草本，高5～10cm；茎基部分枝，斜升或外展；叶肉质，叶片条形至匙形，全缘，无毛；总状花序顶生，花密集；花瓣4，淡紫色或白色带紫色脉纹；长角果圆柱状，略弧曲。花期5～6月。
【分布】内蒙古中部和西部；蒙古。
【生境】荒漠草原至荒漠地带的砂砾质地和干河床。
【用途】饲用。

Thickleaf dontostemon
Mustard family
Perennial herb 5～10 cm tall; stems basally branched, ascending or decumbent; leaves fleshy, blades linear to spatulate, entire, glabrous; racemes terminal, with dense flowers; petals 4, pale purple or white with purple vein-streaks; siliques cylindric, slightly curved. Flowering May to June.
Distribution: C and W Inner Mongolia; Mongolia
Habitat: Sandy and gravelly sites, and dry riverbeds in desert-steppe to desert zones
Use: Forage

扭果花旗杆
Dontostemon elegans Maxim.
十字花科
Brassicaceae

【特征】多年生草本，高15～50cm；茎直立或斜升，具分枝；叶略肉质，条状倒披针形或近匙形，全缘；总状花序顶生；花瓣4，淡紫色，具紫色脉纹；长角果光滑，条状，扭曲。花期5～7月。
【分布】内蒙古西部、甘肃、新疆；蒙古、俄罗斯。
【生境】荒漠地带的洪积平原、戈壁、干河床、冲蚀沟。
【用途】饲用。

Twistedfruit dontostemon
Mustard family
Perennial herb 15～50 cm tall; stems erect or ascending, branched; leaves somewhat fleshy, linear-oblanceolate or subspatulate, entire; racemes terminal; petals 4, pale purple, with purple vein-streaks; siliques smooth, linear, twisted. Flowering May to July.
Distribution: W Inner Mongolia, Gansu and Xinjiang; Mongolia, Russia
Habitat: Alluvial plains, gobi, dry riverbeds and water-eroded ditches in desert zone
Use: Forage

白毛花旗杆
Dontostemon senilis Maxim.
十字花科
Brassicaceae

【特征】多年生草本，高5～15cm，全株被长硬毛；茎直立或斜升，基部呈丛生状分枝；叶条形，宽约1mm，全缘；总状花序顶生；花瓣4，淡紫色或带白色；长角果极细长，长3～4cm，直或稍弧曲。花果期5～10月。
【分布】内蒙古中部和西部、宁夏、甘肃、新疆；蒙古。
【生境】荒漠草原至荒漠地带的石质山坡、干河床。
【用途】饲用。

Oldman dontostemon
Mustard family
Perennial herb 5～15 cm tall, hirsute throughout; stems erect or ascending, basally tufted-branching; leaves linear, about 1 mm wide, entire; racemes terminal; petals 4, pale purple or whitish; siliques very slender, 3～4 cm long, straight or slightly curved. Flowering and fruiting May to October.
Distribution: C and W Inner Mongolia, Ningxia, Gansu and Xinjiang; Mongolia
Habitat: Rocky slopes and dry riverbeds in desert-steppe to desert zones
Use: Forage

拍摄人：赵利清 Photo by Zhao Liqing

拍摄人：赵利清 Photo by Zhao Liqing

扁茎黄耆 (背扁黄耆)
Astragalus complanatus R. Br. ex Bunge
豆科
Fabaceae (Leguminosae)

【特征】多年生草本；茎单一至多数，平卧，长20～100cm，有棱，略扁；单数羽状复叶，小叶9～25，小叶片椭圆形或卵状椭圆形；总状花序腋生，具花3～9枚；蝶形花冠蓝紫色或带白色；柱头有簇毛；荚果纺锤状矩圆形，略膨胀。花期4～6月。
【分布】东北、华北、西北，江苏、河南、四川。
【生境】草甸、灌丛，山坡、沟边。
【用途】饲用；绿肥；种子药用。

Flatstem milkvetch
Pea family
Perennial herb; stems single or many, procumbent, 20～100 cm long, ribbed, slightly compressed; leaves odd-pinnate, leaflets 9～25, blades elliptic or ovate-elliptic; racemes axillary, with 3～9 flowers; papilionaceous corolla blue-purple or whitish; stigmas with a coma; pods fusiform-oblong, slightly inflated. Flowering April to June.
Distribution: NE, N and NW China, Jiangsu, Henan and Sichuan
Habitat: Meadows and thickets, slopes and gully sides
Use: Forage; green manure; seeds for medicine

变异黄耆
Astragalus variabilis Bunge ex Maxim.
豆科
Fabaceae (Leguminosae)

【特征】多年生草本，高10～20cm，全株被伏贴毛，灰绿色；茎直立或稍斜升，由基部分枝；单数羽状复叶，小叶11～15，小叶片条状矩圆形至披针形；总状花序具花5～9枚；蝶形花冠蓝紫色、淡紫色或淡紫红色；荚果条状矩圆形，略弯，密被伏贴毛。花期4～6月。
【分布】内蒙古西部、宁夏、甘肃、青海；蒙古。
【生境】荒漠地带的干河床、低洼地、浅沟。
【 注 】植株含苦马豆素，对牲畜有毒。

Variable milkvetch
Pea family
Perennial herb 10～20 cm tall, appressed-hairy throughout, grey-green; stems erect or slightly ascending, branched from base; leaves odd-pinnate, leaflets 11～15, blades linear-oblong to lanceolate; racemes with 5～9 flowers; papilionaceous corolla bluish-purple, pale purple or pale purplish-red; pods linear-oblong, slightly curved, densely appressed-hairy. Flowering April to June.
Distribution: W Inner Mongolia, Ningxia, Gansu and Qinghai; Mongolia
Habitat: Dry riverbeds, swales and shallow ditches in desert zone
Note: Plants containing swainsonine, poisonous to livestock

玉门黄耆
Astragalus yumenensis S. B. Ho
豆科
Fabaceae (Leguminosae)

【特征】多年生草本，高15～30cm，全株被伏贴毛，灰绿色；茎不发达；单数羽状复叶，小叶5～9，小叶片条形或条状披针形；总状花序具花5～15枚；蝶形花冠淡紫红色或蓝紫色，翼瓣先端微凹；荚果圆柱状，被白色和少量黑色的伏贴毛。花期5～6月。
【分布】内蒙古西部、甘肃西部。
【生境】荒漠地带的砾石质地。
【用途】饲用。

Yumen milkvetch
Pea family
Perennial herb 15～30 cm tall, appressed-hairy throughout, grey-green; stems undeveloped; leaves odd-pinnate, leaflets 5～9, blades linear or linear-lanceolate; racemes with 5～15 flowers; papilionaceous corolla pale purplish-red or bluish-purple, wings retuse at apex; pods cylindric, with white and few black appressed-hairs. Flowering May to June.
Distribution: W Inner Mongolia and W Gansu
Habitat: Gravelly sites in desert zone
Use: Forage

胀果甘草
Glycyrrhiza inflata Batalin
豆科
Fabaceae (Leguminosae)

【特征】多年生草本，高50～150cm；根和根状茎粗壮，内部黄色；茎直立，多分枝；单数羽状复叶，小叶3～9，密被黄褐色腺点，边缘多少波状；总状花序腋生，花疏生；蝶形花冠紫色或淡紫色；荚果椭圆状或矩圆状，被褐色腺点、刺毛状腺体和长柔毛。花期5～7月。
【分布】甘肃、新疆；中亚。
【生境】河岸阶地、沟边、田边、荒地。
【用途】饲用；根和根状茎药用。

Puffedfruit licorice
Pea family
Perennial herb 50～150 cm tall; roots and rhizomes stout, yellow inside; stems erect, much branched; leaves odd-pinnate, leaflets 3～9, densely yellow-brown glandular, margins somewhat undulate; racemes axillary, with loose flowers; papilionaceous corolla purple or pale purple; pods ellipsoid or oblong, brown-glandular, setaceous-glandular and villous. Flowering May to July.
Distribution: Gansu and Xinjiang; Central Asia
Habitat: River terraces, ditch sides, farmland sides and waste places
Use: Forage; roots and rhizomes for medicine

少花米口袋
Gueldenstaedtia verna (Georgi) Boriss.
豆科
Fabaceae (Leguminosae)

【特征】多年生草本，高5～20cm；茎短缩；单数羽状复叶，小叶7～21，小叶片长卵形至披针形，两面被长柔毛，或上面近无毛；伞形花序具花2～4枚；蝶形花冠蓝紫色或紫红色；荚果圆筒状，被长柔毛。花期5月。
【分布】黑龙江、吉林、内蒙古；西伯利亚。
【生境】草原区沙质或砾质地。
【用途】饲用；药用。

Springflower ricepocket
Pea family
Perennial herb 5～20 cm tall; stems shortened; leaves odd-pinnate, leaflets 7～21, blades long-ovate to lanceolate, villous both sides, or subglabrate above; umbels with 2～4 flowers; papilionaceous corolla bluish-purple or purple-red; pods cylindric, villous. Flowering May.
Distribution: Heilongjiang, Jilin and Inner Mongolia; Siberia
Habitat: Sandy and gravelly sites in steppe areas
Use: Forage; medicine

长萼鸡眼草
Kummerowia stipulacea (Maxim.) Makino
豆科
Fabaceae (Leguminosae)

【特征】一年生草本，高5～20cm；茎斜升、斜倚或直立，有分枝，被硬毛；掌状三出复叶，小叶片倒卵形至倒卵状楔形，下面中脉及边缘被硬毛，叶脉明显；花1～3枚，腋生；蝶形花冠淡紫红色。花期7～8月。
【分布】东北、华北、西北、华东、中南；朝鲜、日本、俄罗斯（远东）。
【生境】草原及森林草原区的山地、丘陵、低湿地。
【用途】饲料；绿肥；药用。

Korean lespedeza (Korean clover)
Pea family
Annual herb 5～20 cm tall; stems ascending, decumbent or erect, branched, hispid; leaves palmately trifoliolate, leaflet blades obovate to obovate-cuneate, hispid on midvein below and at margins, veins obvious; flowers 1～3, axillary; papilionaceous corolla pale purplish-red. Flowering July to August.
Distribution: NE, N, NW, E and CS China; Korea, Japan, Russia (Far East)
Habitat: Mountains, hills and moist lowlands in steppe and forest-steppe areas
Use: Forage; green manure; medicine

右图拍摄人：赵利清 Right photo by Zhao Liqing

紫花苜蓿
Medicago sativa L.
豆科
Fabaceae (Leguminosae)

【特征】多年生草本，高30～100cm；茎直立、斜升或平卧，多分枝；羽状三出复叶，小叶边缘1/3以上具锯齿；总状花序短，花5～40枚；蝶形花冠紫色、蓝紫色、淡蓝色、黄绿色、黄色，偶有白色；荚果螺旋状卷曲1～6圈。花期5～7月。

【分布】原产于中亚和伊朗。全国和世界各地广泛栽培。

【生境】个别逸生于草地、田间、路边、宅旁。

【用途】饲用；绿肥；蜜源。

Alfalfa (Lucerne)
Pea family

Perennial herb 30～100 cm tall; stems erect, ascending or procumbent, much branched; leaves pinnately trifoliolate, leaflets marginally serrate in 1/3 above; racemes short, with 5～40 flowers; papilionaceous corolla purple, bluish-purple, pale blue, yellowish-green or yellow, rarely white; pods in spiral of 1～6 turns. Flowering May to July.

Distribution: Native to Central Asia and Iran. Widely cultivated in China and the world

Habitat: Sporadically escaping from cultivation, found in grasslands, fields, roadsides and residential places

Use: Forage; green manure; honey source

镰荚棘豆
Oxytropis falcata **Bunge**
豆科
Fabaceae (Leguminosae)

【特征】多年生草本，高3～35cm，具黏性和异味；茎短缩，木质，丛生；单数羽状复叶，小叶25～45，小叶片条状披针形或条形，下面密被淡褐色腺点；总状花序头状，花6～10枚；蝶形花冠蓝紫色或紫红色，翼瓣先端斜微凹；荚果革质，镰状矩圆形，被腺点和柔毛。花期6～8月。

【分布】西北，四川、西藏；蒙古。

【生境】海拔2 700～5 200m的亚高山和高山带，山坡、河谷、河岸、林下、灌丛、草甸、冰川阶地。

【用途】饲用；药用。

Sicklepod locoweed
Pea family
Perennial herb 3～35 cm tall, viscid and odoriferous; stems shortened, woody, tufted; leaves odd-pinnate, leaflets 25～45, blades linear-lanceolate or linear, densely brownish-glandular below; racemes capitate, with 6～10 flowers; papilionaceous corolla bluish-purple or purple-red, wings obliquely retuse at apex; pods leathery, falcate-oblong, glandular and villous. Flowering June to August.
Distribution: NW China, Sichuan and Tibet; Mongolia
Habitat: Slopes, valleys, river banks, forests, thickets, meadows and glacier terraces in subalpine to alpine zones at 2 700～5 200 m
Use: Forage; medicine

拍摄人：赵利清 Photos by Zhao Liqing

单叶棘豆
Oxytropis monophylla Grubov
豆科
Fabaceae (Leguminosae)

【特征】多年生草本，高3～8cm；无地上茎；单叶，叶柄长5～10mm，叶片披针状椭圆形，全缘；花单生，无梗；花萼密被白色长柔毛；蝶形花冠淡紫色；荚果卵球状，膨胀，密生长柔毛。花期4～6月。
【分布】内蒙古西部、宁夏；蒙古。
【生境】山地荒漠。
【用途】饲用。

Singleleaf locoweed
Pea family
Perennial herb 3～8 cm tall; aculescent; leaves simple, petiole 5～10 mm long, blade lanceolate-elliptic, entire; flowers solitary, sessile; calyx densely white-villous; papilionaceous corolla pale purple; pods ovoid-globose, inflated, densely villous. Flowering April to June.
Distribution: W Inner Mongolia and Ningxia; Mongolia
Habitat: Montane desert
Use: Forage

胶黄耆状棘豆
Oxytropis tragacanthoides Fisch. ex DC.
豆科
Fabaceae (Leguminosae)

【特征】垫状半灌木，高5～30cm；茎短缩，多分枝，呈半球状；单数羽状复叶，叶轴硬化成刺，小叶7～13，密被绢毛；短总状花序，花2～5枚；蝶形花冠蓝紫色或紫红色，稀白色，翼瓣上部极扩展，斜倒卵形；荚果膜质，膨胀，密被柔毛。花期6～8月。
【分布】西北，内蒙古西部；哈萨克斯坦、蒙古、西伯利亚。
【生境】荒漠区海拔2 000～4 100m的山顶、山坡、河谷。
【用途】饲用；观赏。

Tragacanth locoweed
Pea family
Cushion-like subshrub 5～30 cm tall; stems shortened, much branched, forming a hemispheric stand; leaves odd-pinnate, rachis hardened into a prick, leaflets 7～13, densely sericeous; short racemes with 2～5 flowers; papilionaceous corolla blue-purple or purple-red, rarely white, wings strongly extended above, obliquely obovate; pods membranous, inflated, densely pubescent. Flowering June to August.
Distribution: NW China, W Inner Mongolia; Kazakhstan, Mongolia, Siberia
Habitat: Hilltops, slopes and valleys at 2 000～4 100 m in desert areas
Use: Forage; ornamental

粗根老鹳草
Geranium dahuricum DC.
牻牛儿苗科
Geraniaceae

【特征】多年生草本，高20～70cm；根多数，纺锤状；茎直立，被倒向伏毛；叶片掌状5～7深裂几达基部，裂片羽状深裂，小裂片全缘；总花梗具2花；花瓣5，淡紫色或紫红色，或白色带紫色脉纹；蒴果具喙，密被伏毛。花期7～8月。
【分布】东北、华北、西北；蒙古、朝鲜、日本、俄罗斯。
【生境】林下、林缘、灌丛、草甸。
【用途】饲用；药用；根提取栲胶。

Dahurian cranesbill
Geranium family
Perennial herb 20～70 cm tall; roots many, fusiform; stems erect, retrorsely appressed-hairy; leaf blades palmately 5- to 7-parted near base, segments pinnatipartite, lobules entire; flowers 2 per peduncle; petals 5, pale purple or purplish-red, or white with purplish vein-streaks; capsules beaked, densely appressed-hairy. Flowering July to August.
Distribution: NE, N and NW China; Mongolia, Korea, Japan, Russia
Habitat: Forests, forest margins, thickets and meadows
Use: Forage; medicine; roots for extracting tannin

拍摄人：赵利清 Photos by Zhao Liqing

突节老鹳草
Geranium krameri Franch. et Sav.
牻牛儿苗科
Geraniaceae

【特征】多年生草本，高40～100cm；茎直立或稍斜升，关节处略膨大；叶5～7掌状深裂几达基部，裂片5～7深裂，小裂片具缺刻和粗齿；聚伞花序，花2枚；花序轴和花梗被伏毛；花瓣5，粉色或淡紫色，具深色脉纹；蒴果具喙，疏被短柔毛。花期7～8月。

【分布】东北、华北；朝鲜、日本、俄罗斯（远东）。

【生境】草甸、灌丛、林缘。

【用途】饲用。

Tumidnode cranesbill
Geranium family
Perennial herb 40～100 cm tall; stems erect or slightly ascending, joints slightly inflated; leaves palmately 5- to 7-parted near base, segments 5- to 7-parted, lobules notched and coarsely serrate; cymes with 2 flowers; rachis and pedicels appressed-hairy; petals 5, pink or pale purple with dark vein-streaks; capsules beaked, sparsely pubescent. Flowering July to August.

Distribution: NE and N China; Korea, Japan, Russia (Far East)
Habitat: Meadows, thickets and forest edges
Use: Forage

拍摄人：赵 凡 Photos by Zhao Fan

拍摄人：赵利清 Photo by Zhao Liqing

毛蕊老鹳草
Geranium platyanthum Duthie
(*Geranium eriostemon* Fisch. ex DC.)
牻牛儿苗科
Geraniaceae

【特征】多年生草本，高30～80cm；茎直立，被糙毛，上部有腺毛；叶掌状5中裂或略深，裂片菱状卵形，边缘具浅缺刻或粗齿；聚伞花序顶生，花序梗具2～4花；花梗密被腺毛，果期直立；花瓣5，淡蓝紫色，反折；蒴果具喙，被腺毛和糙毛。花期6～8月。
【分布】东北、华北、西北；蒙古、朝鲜、俄罗斯（西伯利亚、远东）。
【生境】林下、林缘、灌丛、草甸。
【用途】饲用；药用；茎、叶提取栲胶。

Broadflower cranesbill
Geranium family
Perennial herb 30～80 cm tall; stems erect, hispidulous, glandular-hairy above; leaves palmately 5-cleft or somewhat parted, segments rhombic-ovate, notched or coarsely serrate; cymes terminal, peduncles with 2～4 flowers; pedicels densely glandular-hairy, erect in fruit; petals 5, pale bluish-purple, reflexed; capsules beaked, glandular-hairy and hispidulous. Flowering June to August.
Distribution: NE, N and NW China; Mongolia, Korea, Russia (Siberia and Far East)
Habitat: Forests, forest margins, thickets and meadows
Use: Forage; medicine; stems and leaves for extracting tannin

拍摄人：赵利清 Photos by Zhao Liqing

野亚麻
Linum stelleroides Planch.
亚麻科
Linaceae

【特征】一年生草本，高40～70cm；茎直立，上部分枝；叶互生，条形至条状披针形；聚伞花序多分枝；萼片5，边缘具黑色腺点；花瓣5，淡紫色至淡蓝紫色；蒴果球形或扁球形。花期6～8月。
【分布】东北、华北、西北、华东；朝鲜、日本、西伯利亚。
【生境】山坡、路旁、荒地。
【用途】饲用；纤维材料；榨油；药用。

Wild flax
Flax family
Annual herb 40～70 cm tall; stems erect, branched above; leaves alternate, linear to linear-lanceolate; cymes much branched; sepals 5, black-glandular at margins; petals 5, pale purple to pale bluish-purple; capsules globose or oblate. Flowering June to August.
Distribution: NE, N, NW and E China; Korea, Japan, Siberia
Habitat: Slopes, roadsides and wastelands
Use: Forage; fiber materials; extracting oil; medicine

新疆花葵
Lavatera cachemiriana Cambess.
锦葵科
Malvaceae

【特征】多年生草本,高达1m,全株被星状绒毛;基生叶近圆形,上部叶3～5裂,裂片三角形,边缘具圆齿;花顶生,近总状排列,或簇生于叶腋;花冠粉紫色,花瓣5,先端2裂。花期6～8月。
【分布】新疆;俄罗斯、中亚、巴基斯坦、印度、尼泊尔。
【生境】湿润草地。
【用途】观赏。

Kashmir treemallow
Mallow family
Perennial herb to 1 m tall, stellate-tomentose throughout; basal leaves suborbicular, upper leaves 3- to 5-lobed, lobes triangular with crenate margins; flowers terminal, nearly racemose-arranged, or fascicled in axils; corolla pinkish-purple, petals 5, apex 2-lobed. Flowering June to August.
Distribution: Xinjiang; Russia, Central Asia, Pakistan, India, Nepal
Habitat: Moist grasslands
Use: Ornamental

拍摄人:卢欣石(上图、中图)&吴新宏(下图)
Photos by Lu Xinshi & Wu Xinhong

球果堇菜
Viola collina Besser
堇菜科
Violaceae

【特征】多年生草本,花期高4~9cm;无地上茎;叶基生,叶片宽卵形或近圆形,边缘具钝齿,两面密被柔毛;花两侧对称;花瓣5,淡紫色或近白色,具紫色脉纹,最下瓣片基部具距,距长约3.5mm;蒴果球形,密被柔毛。花果期5~8月。
【分布】东北、华北、华东、中南、西南;朝鲜、日本、俄罗斯、欧洲。
【生境】林下、林缘、灌丛、草甸、沟谷、路旁。
【用途】饲用;药用。

Hillside violet
Violet family
Perennial herb 4~9 cm tall in flower; acaulescent; leaves basal, blades broadly ovate or suborbicular, blunt-serrate, densely pubescent; flowers zygomorphic; petals 5, pale purple or nearly white with purple streaks, the lower one with a basal spur about 3.5 mm long; capsules globose, densely pubescent. Flowering and fruiting May to August.
Distribution: NE, N, E, CS and SW China; Korea, Japan, Russia, Europe
Habitat: Forests, forest edges, thickets, meadows, gullies and roadsides
Use: Forage; medicine

拍摄人:赵利清 Photo by Zhao Liqing

裂叶堇菜
Viola dissecta Ledeb.
堇菜科
Violaceae

【特征】多年生草本,花期高5~15cm;无地上茎;叶掌状3~5全裂或深裂并再裂,或近羽状深裂,裂片条形;花两侧对称;花瓣5,淡紫色至紫堇色,最下瓣片基部具距,距长5~7mm。花果期5~9月。
【分布】东北、华北、西北;蒙古、朝鲜、俄罗斯(西伯利亚、远东)、中亚。
【生境】山坡、沟谷、林缘、灌丛、草甸。
【用途】饲用;药用。

Cutleaf violet
Violet family
Perennial herb 5~15 cm tall in flower; acaulescent; leaves palmately 3- to 5-divided or parted, further parted or nearly pinnatipartite, segments linear; flowers zygomorphic; petals 5, pale purple to violet, the lower one with a basal spur 5~7 mm long. Flowering and fruiting May to September.
Distribution: NE, N and NW China; Mongolia, Korea, Russia (Siberia and Far East), Central Asia
Habitat: Slopes and gullies, forest edges, thickets and meadows
Use: Forage; medicine

拍摄人：赵利清 Photo by Zhao Liqing

总裂叶堇菜
Viola fissifolia Kitag.
堇菜科
Violaceae

【特征】多年生草本，高6～15cm，全株密被短柔毛；无地上茎；叶片卵形，边缘缺刻状浅裂至中裂，或近羽状深裂，裂片条形；花两侧对称；花瓣5，紫堇色，最下瓣片基部具距，距长6～7mm。花期4～5月。
【分布】东北、华北。
【生境】林缘、灌丛、草甸。
【用途】饲用。

Cleftleaf violet
Violet family
Perennial herb 6～15 cm tall, densely pubescent throughout; acaulescent; leaf blades ovate, margins incised-lobed to cleft, or nearly pinnatipartite, segments linear; flowers zygomorphic; petals 5, violet, the lower one with a basal spur 6～7 mm long. Flowering April to May.
Distribution: NE and N China
Habitat: Forest margins, thickets and meadows
Use: Forage

早开堇菜
Viola prionantha Bunge
堇菜科
Violaceae

【特征】多年生草本,花期高3～10cm;无地上茎;叶多数,基生,叶片矩圆状卵形至狭卵形,边缘具钝齿;花两侧对称;花瓣5,紫堇色或淡紫色,最下瓣片基部具距,距长4～9mm。花果期4～9月。
【分布】东北、华北、西北、华中;朝鲜、俄罗斯(远东)。
【生境】林缘、草地、沟谷、庭院、路边。
【用途】饲用;药用;观赏。

Early violet
Violet family
Perennial herb 3～10 cm tall in flower; acaulescent; leaves numerous, basal, blades oblong-ovate to narrowly ovate, blunt-serrate; flowers zygomorphic; petals 5, violet or pale purple, the lower one with a basal spur 4～9 mm long. Flowering and fruiting April to September.
Distribution: NE, N, NW and C China; Korea, Russia (Far East)
Habitat: Forest edges, grasslands, gullies, courtyards and roadsides
Use: Forage; medicine; ornamental

拍摄人：赵利清 Photo by Zhao Liqing

斑叶堇菜
Viola variegata Fisch. ex Link
堇菜科
Violaceae

【特征】多年生草本，高3~20cm；无地上茎；叶基生，叶片圆形或宽卵形，边缘具圆齿，上面深绿色或绿色，沿脉有白色斑纹，下面通常带紫红色；花两侧对称；花瓣5，红紫色或深紫色，最下瓣片基部具距，距长5~9mm。花果期5~9月。
【分布】东北、华北；朝鲜、日本、俄罗斯（西伯利亚、远东）。
【生境】山坡草地、林下、疏林、灌丛、岩石缝。
【用途】饲用；药用；观赏。

Spottedleaf violet
Violet family
Perennial herb 3~20 cm tall; acaulescent; leaves basal, blades rounded or broadly ovate, crenate, dark green or green with white streaks on veins above, usually purplish-red below; flowers zygomorphic; petals 5, red-purple or dark purple, the lower one with a basal spur 5~9 mm long. Flowering and fruiting May to September.
Distribution: NE and N China; Korea, Japan, Russia (Siberia and Far East)
Habitat: Sloping grasslands, forests, open woodlands, thickets and rock crevices
Use: Forage; medicine; ornamental

箭报春
Primula fistulosa **Turkev.**
报春花科
Primulaceae

【特征】多年生草本；叶基生，莲座状，叶片矩圆形或矩圆状倒披针形，边缘具浅齿；花葶粗壮，管状，高10～17cm；花20枚以上，密集成球状伞形花序；花冠紫红色或带红紫色，高脚碟状，裂片5，先端2深裂。花期5～6月。
【分布】东北，内蒙古东部；蒙古、俄罗斯（远东）。
【生境】低湿地和砂质草甸。
【用途】饲用；观赏。

Fistulous primrose
Primula family
Perennial herb; leaves basal, rosulate, blades oblong or oblong-oblanceolate, shallowly serrate; scapes stout, fistulose, 10～17 cm tall; flowers more than 20, clustered into a globose umbel; corolla purple-red or reddish-purple, salverform, with 5 lobes, lobes 2-parted at apex. Flowering May to June.
Distribution: NE China, E Inner Mongolia; Mongolia, Russia (Far East)
Habitat: Wet lowlands and sandy meadows
Use: Forage; ornamental

拍摄人：赵利清 Photo by Zhao Liqing

耳叶补血草
Limonium otolepis (Schrenk) Kuntze
白花丹科
Plumbaginaceae

【特征】多年生草本，高30～120cm；基生叶倒卵状匙形，花序轴下部节上的叶圆肾形，抱茎；花序圆锥状，多回分枝，下方分枝形成繁多的不育小枝；花萼倒圆锥状，干膜质，萼檐白色；花冠5裂，淡蓝紫色。花期6～7月。
【分布】甘肃、新疆；中亚。
【生境】荒漠地带的沙质盐碱地。
【用途】观赏；药用。

Saltmarsh sea lavender
Leadwort family
Perennial herb 30～120 cm tall; basal leaves obovate-spatulate, nodal leaves below inflorescence rachis rounded-reniform and clasping; inflorescence paniculate, branching several times, lower branches with numerous and sterile branchlets; calyx obconic, scarious, limbs white; corolla 5-lobed, pale bluish-purple. Flowering June to July.
Distribution: Gansu and Xinjiang; Central Asia
Habitat: Sandy saline-alkali sites in desert zone
Use: Ornamental: medicine

拍摄人：赵利清 Photos by Zhao Liqing

椭圆叶花锚
Halenia elliptica D. Don
龙胆科
Gentianaceae

【特征】一年生草本，高15~30cm；茎直立，近四棱，有分枝；叶对生，椭圆形或卵形；聚伞花序顶生或腋生；花冠锚状，蓝色或蓝紫色，4裂片，每一裂片基部具距。花期7~8月。

【分布】西北、西南、华中、内蒙古、山西；尼泊尔、锡金、不丹、印度、中亚。

【生境】海拔700~4 100m的林下、林缘、草地、灌丛、沟谷水边。

【用途】药用；观赏。

Ellipticleaf spurred gentian
Gentian family
Annual herb 15~30 cm tall; stems erect, subquadrangular, branched; leaves opposite, elliptic or ovate; cymes terminal or axillary; corolla anchor-like, blue or bluish-purple, 4-lobed, each lobe spurred at base. Flowering July to August.
Distribution: NW, SW and C China, Inner Mongolia and Shanxi; Nepal, Sikkim, Bhutan, India, Central Asia
Habitat: Forests, forest edges, grasslands, thickets and gully watersides at 700~4 100 m
Use: Medicine; ornamental

羊角子草
Cynanchum cathayense Tsiang et H. D. Zhang
萝藦科
Asclepiadaceae

【特征】多年生藤本，具乳汁；茎缠绕；叶矩圆状戟形或三角状戟形，基部心状戟形，两耳圆形；伞状或伞房状聚伞花序腋生；花冠辐状，5深裂，淡紫色；副花冠单轮；蓇葖果单生，披针状或条状。花期6～8月。
【分布】河北西部、内蒙古西部、宁夏、甘肃、新疆。
【生境】水边及荒漠绿洲、干湖盆、湿润沙地。
【用途】饲用；幼果可食。

Spearleaf swallow-wort
Milkweed family
Perennial vine, with milky juice; stems twining; leaves oblong-hastate or deltoid-hastate, base cordate-hastate with 2 rounded auricles; umbelliform or corymbiform cymes axillary; corolla rotate, 5-parted, pale purple; corona uniseriate; follicle single, lanceolate or linear. Flowering June to August.
Distribution: W Hebei, W Inner Mongolia, Ningxia, Gansu and Xinjiang
Habitat: Watersides, and oasis, dry lake basins and moist sands in desert zone
Use: Forage; young fruits edible

戟叶鹅绒藤
Cynanchum sibiricum Willd.
萝藦科
Asclepiadaceae

【特征】多年生藤本，具乳汁；茎缠绕；叶长戟形或戟状心形，基部具2圆耳；伞房状聚伞花序腋生；花冠辐状，5深裂，外面白色，内面紫色或淡紫色；副花冠双轮；蓇葖果单生，狭披针状。花期5～8月。
【分布】新疆、甘肃；蒙古、俄罗斯、中亚。
【生境】荒漠地带的绿洲及其边缘。
【用途】药用。

Siberian swallow-wort
Milkweed family
Perennial vine, with milky juice; stems twining; leaves long-hastate or hastate-cordate, with 2 rounded auricles at base; corymbiform cymes axillary; corolla rotate, 5-parted, white outer side, purple or pale purple inside; corona biseriate; follicle single, narrowly lanceolate. Flowering May to August.
Distribution: Xinjiang and Gansu; Mongolia, Russia, Central Asia
Habitat: Oasis and its edges in desert zone
Use: Medicine

拍摄人：拾 涛 Photos by Shi Tao

灰毛软紫草
Arnebia fimbriata Maxim.
紫草科
Boraginaceae

【特征】多年生草本，高10~18cm，全株密被长硬毛；茎多分枝；叶矩圆状披针形或狭披针形；聚伞花序，具花2~5枚；花冠淡蓝紫色、粉色或白色，5裂，裂片边缘具齿；小坚果密生瘤状凸起。花期5~6月。
【分布】内蒙古西部、宁夏、甘肃、青海；蒙古。
【生境】荒漠地带的戈壁、沙地、砾石质坡地、干河谷。
【用途】观赏。

Fringed arnebia
Borage family
Perennial herb 10~18 cm tall, densely hirsute throughout; stems much branched; leaves oblong-lanceolate to narrowly lanceolate; cymes with 2~5 flowers; corolla pale bluish-purple, pink or white, 5-lobed with toothed margins; nutlets densely tuberculate. Flowering May to June.
Distribution: W Inner Mongolia, Ningxia, Gansu and Qinghai; Mongolia
Habitat: Gobi, sands, gravelly slopes and dry valleys in desert
Use: Ornamental

荆条
Vitex negundo var. *heterophylla* (Franch.) Rehder
马鞭草科
Verbenaceae

【特征】灌木，高1～2m；小枝四棱形，密被绒毛；掌状复叶，小叶5，有时3，小叶片矩圆状卵形至披针形，浅裂至深裂，边缘有缺刻状锯齿，下面被绒毛；圆锥状聚伞花序顶生；花冠蓝色或淡紫色，二唇形，裂片5；核果近球状。花期7～8月。

【分布】华东、华北、华中；日本。

【生境】山地阳坡、林缘、河谷。

【用途】药用；造纸；编织；水土保持；蜜源；提取芳香油。

Chinese chastetree
Vervain family
Shrub 1～2 m tall; branchlets quadrangular, densely tomentose; leaves palmate, leaflets 5, sometimes 3, blades oblong-ovate to lanceolate, lobed to parted, incised-serrate, tomentose below; paniculate cymes terminal; bilabiate corolla blue or pale purple, 5-lobed; drupes subglobose. Flowering July to August.
Distribution: E, N and C China; Japan
Habitat: Mountain sunny slopes, forest margins and valleys
Use: Medicine; papermaking; weaving; soil conservation; honey source; extracting essential oil

拍摄人：赵利清 Photo by Zhao Liqing

灌木青兰
Dracocephalum fruticulosum Stephan ex Willd.
(沙地青兰 *Dracocephalum psammophilum* C. Y. Wu et W. T. Wang)
唇形科
Lamiaceae (Labiatae)

【特征】小半灌木，高达20cm；茎直立或近直立，密被倒向毛；叶多少肉质，椭圆形至卵状椭圆形，两面密被短毛和腺点；轮伞花序顶生；花冠蓝紫色或淡紫色，二唇形，上唇2浅裂，下唇3裂。花期8月。
【分布】内蒙古西部、宁夏（贺兰山）；蒙古。
【生境】荒漠石质山坡。
【用途】饲用。

Bush dragonhead
Mint family
Shrublet to 20 cm tall; stems erect or suberect, densely retrorse-hairy; leaves somewhat fleshy, elliptic to ovate-elliptic, densely minute-hairy and glandular; verticillasters terminal; bilabiate corolla bluish-purple or pale purple, upper lip 2-lobed, lower lip 3-lobed. Flowering August.
Distribution: W Inner Mongolia and Ningxia (Helan Mountains); Mongolia
Habitat: Rocky slopes in desert
Use: Forage

硬尖神香草
Hyssopus cuspidatus Boriss.
唇形科
Lamiaceae (Labiatae)

【特征】半灌木，高15～60cm；茎四棱，紫红色，由基部多分枝，光滑无毛；叶单生或簇生，条形，全缘，先端具芒状刺，两面具腺点；穗状花序顶生；花萼紫红色，萼齿先端有刺尖；花冠蓝紫色，二唇形，外面疏被柔毛和腺点，上唇直立，2浅裂，下唇3裂。花期6月。

【分布】新疆；中亚。

【生境】山地草原中的砾石山坡。

【用途】药用；观赏。

Prickly hyssop
Mint family
Subshrub 15～60 cm tall; stems quadrangular, purple-red, much branched from base, glabrous; leaves single or fascicled, linear, entire, apex with aristiform spine, glandular both sides; spikes terminal; calyx purple-red, teeth spinose-pointed; bilabiate corolla blue-purple, sparsely pubescent and glandular outside, upper lip erect, 2-lobed, lower lip 3-lobed. Flowering June.

Distribution: Xinjiang; Central Asia
Habitat: Gravelly slopes in montane steppe
Use: Medicine; ornamental

拍摄人：张洪江 Photos by Zhang Hongjiang

尖齿糙苏
Phlomis dentosa Franch.
唇形科
Lamiaceae (Labiatae)

【特征】多年生草本，高20~40cm；茎直立，多分枝，被刚毛和星状毛；叶三角形或三角状卵形，边缘具圆齿；轮伞花序，花多数；花冠粉红色或淡紫色，二唇形，上唇盔状，外面密被星状柔毛，边缘具小齿，下唇3圆裂；小坚果顶端光滑无毛。花期6~8月。
【分布】河北、内蒙古、甘肃、青海。
【生境】山地、沟谷，草甸、草甸草原。
【用途】药用。

Sharptooth Jerusalem sage
Mint family
Perennial herb 20~40 cm tall; stems erect, much branched, setose and stelipilous; leaves triangular or triangular-ovate, crenate; verticillasters with many flowers; bilabiate corolla pink or pale purple, upper lip galeate, stellate-pubescent outside, margins denticulate, lower lip 3-lobed; nutlets apically glabrous. Flowering June to August.
Distribution: Hebei, Inner Mongolia, Gansu and Qinghai
Habitat: Mountains and ravines, meadows and meadow-steppe
Use: Medicine

小裂叶荆芥
Schizonepeta annua (Pall.) Schischk.
唇形科
Lamiaceae (Labiatae)

【特征】一年生草本,高30～40cm;茎直立或斜升,基部多分枝,被柔毛;叶1～2回羽状深裂,两面疏被柔毛和黄色腺点;多数轮伞花序组成穗状花序,顶生;花冠蓝紫色至淡紫色,外面被长柔毛,二唇形,上唇2裂,下唇3裂。花期7～8月。

【分布】内蒙古西部、新疆;蒙古、俄罗斯。

【生境】荒漠地区的丘陵、坡地。

【用途】饲用。

Little schizonepeta
Mint family
Annual herb 30～40 cm tall; stems erect or ascending, much branched at base, pubescent; leaves pinnately to bipinnately parted, sparsely pubescent and yellow-glandular; many verticillasters forming a terminal spike; bilabiate corolla blue-purple to pale purple, villous outside, upper lip 2-lobed, lower lip 3-lobed. Flowering July to August.
Distribution: W Inner Mongolia and Xinjiang; Mongolia, Russia
Habitat: Hills and hillsides in desert areas
Use: Forage

长腺小米草
Euphrasia hirtella Jord. ex Reuter
玄参科
Scrophulariaceae

【特征】一年生草本，高3~40cm；茎直立，单一，细弱，被柔毛；茎下部叶较小，向上渐大，边缘具齿；苞叶叶状；叶及苞叶被硬毛和长腺毛；穗状花序顶生；花冠白色或淡紫色，二唇形，上唇直立，2浅裂，下唇开展，3裂。花期6~8月。

【分布】东北、内蒙古、新疆；蒙古、朝鲜、俄罗斯、欧洲。

【生境】草甸、草原、林地、林缘。

【用途】饲用。

Smallflower sticky eyebright
Figwort family
Annual herb 3~40 cm tall; stems erect, single, slender, pubescent; lower cauline leaves smaller, gradually larger upward, margins toothed; bracteal leaves leaf-like; leaves and bracteal leaves hispidulous and long-piloglandulose; spikes terminal; bilabiate corolla white or pale purple, upper lip erect, 2-lobed, lower lip spreading, 3-lobed. Flowering June to August.
Distribution: NE China, Inner Mongolia and Xinjiang; Mongolia, Korea, Russia, Europe
Habitat: Meadows, steppe, forests and forest edges
Use: Forage

拍摄人：赵利清（左图）& 乔 江（右图） Photos by Zhao Liqing & Qiao Jiang

小米草
Euphrasia pectinata Ten.
玄参科
Scrophulariaceae

【特征】一年生草本，高10～30cm；茎直立，常单一，被柔毛；茎下部叶较小，向上渐大，边缘具齿；苞叶叶状；叶及苞叶被短硬毛；穗状花序顶生；花冠白色或淡紫色，二唇形，上唇直立，2浅裂，下唇开展，3裂。花期7～8月。
【分布】东北、华北、西北；蒙古、日本、俄罗斯、欧洲。
【生境】山地草甸、草原、林缘、灌丛。
【用途】饲用；药用。

Common eyebright
Figwort family
Annual herb 10～30 cm tall; stems erect, usually single, pubescent; lower cauline leaves smaller, gradually larger upward, margins toothed; bracteal leaves leaf-like; leaves and bracteal leaves hispidulous; spikes terminal; bilabiate corolla white or pale purple, upper lip erect, 2-lobed, lower lip spreading, 3-lobed. Flowering July to August.
Distribution: NE, N and NW China; Mongolia, Japan, Russia, Europe
Habitat: Montane meadows, steppe, forest edges and thickets
Use: Forage; medicine

北水苦荬
Veronica anagallis-aquatica L.
玄参科
Scrophulariaceae

【特征】多年生水生草本,稀一年生,高10~100cm;茎直立或基部倾斜,单一或有分枝;上部叶半抱茎,叶片椭圆形或长卵形,全缘或有疏齿;总状花序腋生;花梗斜升,与花序轴呈锐角;花冠辐状,4深裂,淡蓝色、淡紫色或白色。花期7~9月。

【分布】长江以北及西南;蒙古、朝鲜、俄罗斯、中亚、巴基斯坦、尼泊尔、欧洲、北美洲。

【生境】溪水边、沼泽。

【用途】药用。

Water speedwell
Figwort family
Perennial aquatic herb, rarely annual, 10~100 cm tall; stems erect or decumbent at base, single or branched; upper leaves clasping, blades elliptic or long-ovate, entire or spaced-serrate; racemes axillary; pedicels ascending, at acute angle with rachis; corolla rotate, 4-parted, pale blue, pale purple or white. Flowering July to September.

Distribution: Regions to north of Changjiang River, and SW China; Mongolia, Korea, Russia, Central Asia, Pakistan, Nepal, Europe, North America

Habitat: Streamsides and swamps

Use: Medicine

拍摄人：赵利清 Photos by Zhao Liqing

草本威灵仙 (轮叶婆婆纳)
Veronicastrum sibiricum (L.) Pennell
玄参科
Scrophulariaceae

【特征】多年生草本，高约100cm；茎直立，单一，不分枝；叶4～6枚轮生，叶片矩圆状披针形至披针形，边缘具锯齿；总状花序顶生；花萼5深裂；花冠筒状，4裂，红紫色至淡紫色。花期7～9月。
【分布】东北、华北；朝鲜、日本、俄罗斯（西伯利亚、远东）。
【生境】山地林下、林缘、灌丛、草甸。
【用途】药用。

Siberian speedwell
Figwort family
Perennial herb about 100 cm tall; stems erect, single, unbranched; leaves in whorls of 4～6, blades oblong-lanceolate to lanceolate, serrate; racemes terminal; calyx 5-parted; corolla tubular, 4-lobed, reddish-purple to pale purple. Flowering July to September.
Distribution: NE and N China; Korea, Japan, Russia (Siberia and Far East)
Habitat: Montane forests, forest margins, thickets and meadows
Use: Medicine

拍摄人：赵利清 Photos by Zhao Liqing

宁夏沙参
Adenophora ningxianica Hong
桔梗科
Campanulaceae

【特征】多年生草本，高10～50cm，有白色乳汁；茎丛生，不分枝；茎生叶互生，披针形，边缘有锯齿；假总状花序顶生或腋生，无分枝，花下垂；萼裂片具1对小齿或全缘；花冠蓝紫色至淡紫色，狭钟状，5浅裂；花柱稍长于花冠。花期7～8月。
【分布】内蒙古、宁夏、甘肃。
【生境】荒漠带山地阴坡岩石缝。
【用途】饲用。

Ningxia ladybells
Bellflower family
Perennial herb 10～50 cm tall, with white milky juice; stems tufted, unbranched; cauline leaves alternate, lanceolate, serrate; pseudo-racemes terminal or axillary, unbranched, flowers nodding; sepals toothed in 1 pairs or entire; corolla blue-purple to pale purple, narrowly campanulate, 5-lobed; style slightly longer than the corolla. Flowering July to August.
Distribution: Inner Mongolia, Ningxia and Gansu
Habitat: Rock crevices on mountain shady slopes in desert zone
Use: Forage

三脉紫菀
Aster ageratoides Turcz.
菊科
Asteraceae (Compositae)

【特征】多年生草本，高40～60cm；茎直立，单一，常带紫色，上部稍分枝；叶卵形至披针形，具离基三出脉，边缘具锯齿，两面被短硬毛和腺点；头状花序伞房状排列；舌状小花紫色、粉色或白色，管状小花黄色。花期8～9月。
【分布】几遍全国；朝鲜、日本、俄罗斯（西伯利亚、远东）。
【生境】山地和丘陵，林缘和草地。
【用途】饲用；药用。

Whiteweed aster
Aster family
Perennial herb 40～60 cm tall; stems erect, single, usually purplish, slightly branched above; leaves ovate to lanceolate, triple-nerved, serrate, hispidulous and glandular; heads corymbiform-arranged; ray florets purple, pink or white, disk florets yellow. Flowering August to September.
Distribution: Almost throughout China; Korea, Japan, Russia (Siberia and Far East)
Habitat: Mountains and hills, forest margins and grasslands
Use: Forage; medicine

拍摄人：赵利清 Photo by Zhao Liqing

火烙草
Echinops przewalskii Iljin
菊科
Asteraceae (Compositae)

【特征】多年生草本，高30～40cm；茎直立，密被白绵毛；叶革质，2回羽状深裂至半裂，裂片扭曲，具针刺，下面密被灰白色绵毛，上部叶变小；复头状花序球形，单生茎顶，灰蓝色或淡紫色。花期6～8月。
【分布】内蒙古、山西、甘肃、山东；蒙古。
【生境】石质或石砾质山坡、戈壁。

Flaming globethistle
Aster family
Perennial herb 30～40 cm tall; stems erect, densely white-lanate; leaves leathery, bipinnately parted to cleft, segments twisted with acicular spines, densely grey-white lanate below, upper leaves reduced; compound heads globose, solitary and terminal, grey-blue or pale purple. Flowering June to August.
Distribution: Inner Mongolia, Shanxi, Gansu and Shandong; Mongolia
Habitat: Rocky or gravelly slopes and gobi

拍摄人：赵利清 Photos by Zhao Liqing

飞蓬
***Erigeron acer* L.**
菊科
Asteraceae (Compositae)

【特征】二年生草本，高10~60cm；茎直立，单一，具纵棱，被柔毛和硬毛；叶两面被硬毛，全缘或有齿；头状花序排列成伞房状或圆锥状；总苞片密被硬毛；舌状小花淡紫色，管状小花黄色。花期7~8月。
【分布】中国北方大部分省区；蒙古、日本、西伯利亚、中亚、欧洲、北美洲。
【生境】山坡、林缘、草甸、河岸、田边。
【用途】饲用；药用。

Bitter fleabane
Aster family
Biennial herb 10~60 cm tall; stems erect, single, ribbed, pubescent and hirsute; leaves hirsute, entire or toothed; heads corymbiform- or paniculate-arranged; involucral bracts densely hirsute; ray florets pale purple, disk florets yellow. Flowering July to August.
Distribution: Most regions of northern China; Mongolia, Japan, Siberia, Central Asia, Europe, North America
Habitat: Slopes, forest edges, meadows, river banks and farmland sides
Use: Forage; medicine

山莴苣
Lagedium sibiricum (L.) Soják
(*Lactuca sibirica* (L.) Benth. ex Maxim.)
菊科
Asteraceae (Compositae)

【特征】多年生草本，高20～90cm，含白色乳汁；茎直立，上部有分枝；叶披针形至条状披针形，全缘或有浅齿或缺刻；头状花序排列成伞房状或圆锥状；总苞片紫红色；舌状小花蓝紫色或淡紫色。花期7～8月。
【分布】东北、华北、西北；蒙古、朝鲜、日本、俄罗斯（西伯利亚、远东）、中亚、欧洲、北美洲。
【生境】林下、林缘、草甸、田间、路旁。
【用途】饲用；药用。

Siberian lettuce
Aster family
Perennial herb 20～90 cm tall, with white milky juice; stems erect, branched above; leaves lanceolate to linear-lanceolate, entire or shallowly toothed or incised; heads corymbose- or paniculate-arranged; involucral bracts purple-red; ligulate florets blue-purple or pale purple. Flowering July to August.
Distribution: NE, N and NW China; Mongolia, Korea, Japan, Russia (Siberia and Far East), Central Asia, Europe, North America
Habitat: Forests, forest margins, meadows, farmlands and roadsides
Use: Forage; medicine

乳苣
Mulgedium tataricum (L.) DC.
(蒙山莴苣 *Lactuca tatarica* (L.) C. A. Mey.)
菊科
Asteraceae (Compositae)

【特征】多年生草本，高10～70cm，含白色乳汁；茎直立；茎下部叶羽状或倒向羽状深裂至浅裂，裂片边缘具刺齿，上部叶全缘；头状花序多数，圆锥状排列；总苞片紫红色；舌状小花蓝紫色或淡紫色。花期6～7月。
【分布】东北至西北；蒙古、西伯利亚、中亚、印度、伊朗、欧洲。
【生境】草甸、固定沙丘、田间、路旁。
【用途】饲用；食用；药用。

Blue lettuce
Aster family
Perennial herb 10～70 cm tall, with milky juice; stems erect; lower cauline leaves pinnately or runcinately parted to lobed, lobe margins with spinose teeth, upper leaves entire; many heads paniculate-arranged; involucral bracts purple-red; ligulate florets blue-purple or pale purple. Flowering June to July.
Distribution: NE to NW China; Mongolia, Siberia, Central Asia, India, Iran, Europe
Habitat: Meadows, fixed dunes, farmlands and roadsides
Use: Forage; edible; medicine

华北乌头
Aconitum jeholense var. *angustius* (W. T. Wang) Y. Z. Zhao
毛茛科
Ranunculaceae

【特征】多年生草本，高70～120cm；茎直立，粗壮；叶掌状3全裂，裂片细裂，小裂片狭条形或条形，两面无毛；总状花序顶生，长10～40cm，花10～35枚；花序轴和花梗疏被柔毛或近无毛；萼片5，蓝紫色，上萼片浅盔状。花期8月。
【分布】华北，山东；蒙古、俄罗斯（西伯利亚、远东）。
【生境】山地林缘、灌丛、草甸。
【用途】有毒植物。观赏。

North China monkshood
Buttercup family
Perennial herb 70～120 cm tall; stems erect, stout; leaves palmately 3-divided, segments dissected, lobules narrowly linear or linear, glabrous; racemes terminal, 10～40 cm long, with flowers 10～35; rachis and pedicels sparsely pubescent or subglabrate; sepals 5, blue-purple, upper sepal shallowly galeate. Flowering August.
Distribution: N China, Shandong; Mongolia, Russia (Siberia and Far East)
Habitat: Montane forest margins, thickets and meadows
Use: Poisonous. Ornamental

阴山乌头
Aconitum yinschanicum Y. Z. Zhao
毛茛科
Ranunculaceae

【特征】多年生草本，高50～100cm；茎直立，疏被短柔毛；叶掌状3全裂，裂片细裂，小裂片条形；总状花序顶生，下部有时分枝，具多花；花序轴和花梗密被柔毛；萼片5，蓝紫色，外面被柔毛，上萼片盔状。花期8月。

【分布】内蒙古（阴山）。

【生境】山地灌丛和草甸，沟谷边缘。

【用途】有毒植物。药用；观赏。

Yinshan monkshood
Buttercup family
Perennial herb 50～100 cm tall; stems erect, sparsely pubescent; leaves palmately 3-divided, segments dissected, lobules linear; racemes terminal, sometimes branched below, with numerous flowers; rachis and pedicels densely pubescent; sepals 5, blue-purple, pubescent outer side, upper sepal galeate. Flowering August.
Distribution: Inner Mongolia (Yinshan Mountains)
Habitat: Montane scrublands and meadows, ravine sides
Use: Poisonous. Medicine; ornamental

右图拍摄人：赵利清 Right photo by Zhao Liqing

左图拍摄人：赵利清 Left photo by Zhao Liqing

白蓝翠雀花
Delphinium albocoeruleum Maxim.
毛茛科
Ranunculaceae

【特征】多年生草本，高10～60cm；茎直立；叶掌状3中裂至全裂，两面被短柔毛；伞房花序具2～7花，稀1花；萼片5，花瓣状，蓝紫色或蓝白色，上萼片基部具距；退化雄蕊黑褐色，腹面具黄色髯毛。花期7～8月。
【分布】内蒙古（贺兰山）、宁夏、甘肃、青海、西藏。
【生境】云杉林缘草甸。
【用途】观赏；药用。

Bluewhite larkspur
Buttercup family
Perennial herb 10～60 cm tall; stems erect; leaves palmately 3-cleft to divided, pubescent; corymbs with flowers 2～7, rarely 1; sepals 5, petal-like, blue-purple or blue-white, the upper sepal basally spurred; staminodia black-brown, ventral surface yellow-bearded. Flowering July to August.
Distribution: Inner Mongolia (Helan Mountains), Ningxia, Gansu, Qinghai and Tibet
Habitat: Meadows along spruce forest edges
Use: Ornamental; medicine

宿根亚麻
Linum perenne L.
亚麻科
Linaceae

【特征】多年生草本，高20～70cm；茎丛生，直立或斜升，由基部分枝；叶互生，条形至条状披针形；聚伞花序，花通常多数；萼片5；花瓣5，蓝色或淡蓝紫色；蒴果近球形。花期6～8月。
【分布】东北、华北、西北、西南；蒙古、俄罗斯、西亚、欧洲。
【生境】砂砾质草地。
【用途】饲用；纤维材料；榨油。

Blue flax
Flax family
Perennial herb 20～70 cm tall; stems tufted, erect or ascending, branched from base; leaves alternate, linear to linear-lanceolate; cymes usually with many flowers; sepals 5; petals 5, blue or pale bluish-purple; capsules subglobose. Flowering June to August.
Distribution: NE, N, NW and SW China; Mongolia, Russia, W Asia, Europe
Habitat: Gravelly grasslands
Use: Forage; fiber materials; extracting oil

左图拍摄人：赵利清 Left photo by Zhao Liqing

拍摄人：赵 凡 Photos by Zhao Fan

秦艽
Gentiana macrophylla Pall.
龙胆科
Gentianaceae

【特征】多年生草本，高30～60cm；茎单一或少数，斜升或直立；基生叶莲座状，狭披针形至狭倒披针形，茎生叶披针形；聚伞花序，花数枚或多数簇生于茎顶，呈头状，或腋生为轮状；花萼膜质，一侧开裂，萼齿长0.5～1mm；花冠蓝色或蓝紫色，管钟状，裂片5。花期7～8月。

【分布】东北、华北、西北、四川；蒙古、俄罗斯（西伯利亚、远东）。

【生境】山地草甸、林缘、灌丛、沟谷。

【用途】药用。

Bigleaf gentian
Gentian family
Perennial herb 30～60 cm tall; stems single or few, ascending or erect; basal leaves rosulate, narrowly lanceolate to narrowly oblanceolate, cauline leaves lanceolate; cymes with several or many flowers, terminal into a capitate cluster, or in axillary whorls; calyx membranous, split one side, teeth 0.5～1 mm long; corolla blue or bluish-purple, tubular-campanulate, 5-lobed. Flowering July to August.

Distribution: NE, N and NW China, and Sichuan; Mongolia, Russia (Siberia and Far East)
Habitat: Montane meadows, forest edges, thickets and gullies
Use: Medicine

龙胆
Gentiana scabra Bunge
龙胆科
Gentianaceae

【特征】多年生草本,高30~60cm;茎直立,常单一;无莲座状基生叶,叶卵形或卵状披针形,叶缘粗糙;花1至数枚簇生于枝顶或上部叶腋;花萼管钟状,萼齿长8~15mm;花冠蓝色或蓝紫色,管钟状,裂片5,先端尖。花期8~9月。
【分布】东北,内蒙古东部、浙江、福建;朝鲜、日本、俄罗斯(西伯利亚、远东)。
【生境】山地林缘、灌丛、草甸。
【用途】药用;观赏。

Japanese gentian
Gentian family
Perennial herb 30~60 cm tall; stems erect, usually single; without rosulate basal leaves, leaves ovate or ovate-lanceolate, with scabrous margins; flowers 1 or several, terminal or also axillary in upper nodes; calyx tubular-campanulate, teeth 8~15 mm long; corolla blue or blue-purple, tubular-campanulate, 5-lobed, apex pointed. Flowering August to September.
Distribution: NE China, E Inner Mongolia, Zhejiang and Fujian; Korea, Japan, Russia (Siberia and Far East)
Habitat: Montane forest edges, thickets and meadows
Use: Medicine; ornamental

管花秦艽
Gentiana siphonantha Maxim. ex Kusn.
龙胆科
Gentianaceae

【特征】多年生草本，高10～30cm；茎少数，直立；基生叶具柄，叶片条形至条状披针形；花多数，簇生于茎顶呈头状，少数花生于上部叶腋；花萼小，长为花冠的1/5至1/4，萼筒一侧开裂或不裂，萼齿长1～3mm；花冠蓝色，管钟状，裂片5。花期7～9月。
【分布】宁夏、甘肃、青海、四川。
【生境】海拔1 800～4 500m的草原、草甸、灌丛。
【用途】药用；观赏。

Tubeflower gentian
Gentian family
Perennial herb 10～30 cm tall; stems few, erect; basal leaves petiolate, leaf blades linear to linear-lanceolate; many flowers terminal into a capitate cluster, and also few axillary in upper nodes; calyx small, 1/5 to 1/4 as long as the corolla, tube split one side or not split, teeth 1～3 mm long; corolla blue, tubular-campanulate, 5-lobed. Flowering July to September.
Distribution: Ningxia, Gansu, Qinghai and Sichuan
Habitat: Steppe, meadows and thickets at 1 800～4 500 m
Use: Medicine; ornamental

三花龙胆
Gentiana triflora Pall.
龙胆科
Gentianaceae

【特征】多年生草本，高30～60cm；茎直立，单一；无莲座状基生叶，叶条状披针形，稀披针形，叶缘光滑；花1～5枚簇生于枝顶及上部叶腋；花萼管钟状，萼筒常一侧浅裂，萼齿长4～8mm；花冠蓝色，管钟状，裂片5，先端钝或圆。花期8～9月。
【分布】东北，内蒙古东部；朝鲜、日本、俄罗斯（西伯利亚、远东）。
【生境】山地林缘、灌丛、草甸。
【用途】药用；观赏。

Threeflower gentian
Gentian family
Perennial herb 30～60 cm tall; stems erect, single; without rosulate basal leaves, leaves linear-lanceolate, rarely lanceolate, with smooth margins; flowers 1～5, terminal and also axillary in upper nodes; calyx tubular-campanulate, tube usually shallowly split one side, teeth 4～8 mm long; corolla blue, tubular-campanulate, 5-lobed, apex obtuse or rounded. Flowering August to September.
Distribution: NE China, E Inner Mongolia; Korea, Japan, Russia (Siberia and Far East)
Habitat: Montane forest margins, thickets and meadows
Use: Medicine; ornamental

拍摄人：赵利清 Photo by Zhao Liqing

拍摄人：赵 凡 Photos by Zhao Fan

花荵
Polemonium caeruleum L.
花荵科
Polemoniaceae

【特征】多年生草本，高40～80cm；茎单一，不分枝，上部被腺毛；单数羽状复叶，小叶互生，卵状披针形至披针形，全缘；聚伞圆锥花序具多花；花冠蓝色或蓝紫色，钟状，5裂，边缘稀具睫毛。花期6～7月。
【分布】东北、华北、新疆、云南；亚洲、欧洲、北美洲。
【生境】疏林、灌丛、草甸、溪谷。
【用途】观赏；药用。

Greek valerian
Polemonium family
Perennial herb 40～80 cm tall; stems single, unbranched, glandular-pubescent above; leaves odd-pinnate, leaflets alternate, ovate-lanceolate to lanceolate, entire; thyrse with many flowers; corolla blue or bluish-purple, campanulate, 5-lobed, margins rarely ciliate. Flowering June to July.
Distribution: NE and N China, Xinjiang and Yunnan; Asia, Europe, North America
Habitat: Open woodlands, scrublands, meadows and gullies
Use: Ornamental; medicine

蓝蓟
Echium vulgare L.
紫草科
Boraginaceae

【特征】二年生草本，高60~100cm，全株密被糙硬毛；茎有分枝；叶条状披针形至披针形；聚伞花序狭长，花多数；花冠蓝紫色，斜钟状，两侧对称，5浅裂，上方裂片较大，外面被毛。花期7~8月。
【分布】新疆北部；亚洲西部至欧洲。
【生境】山地草原。
【用途】观赏。

Blueweed (Blue-thistle)
Borage family
Biennial herb 60~100 cm tall, densely hispid throughout; stems branched; leaves linear-lanceolate to lanceolate; cymes narrow and long, with numerous flowers; corolla blue-purple, obliquely campanulate, zygomorphic, 5-lobed, the upper lobe larger with hairs outside. Flowering July to August.
Distribution: N Xinjiang; W Asia to Europe
Habitat: Montane steppe
Use: Ornamental

北齿缘草
Eritrichium borealisinense Kitag.
紫草科
Boraginaceae

【特征】多年生草本，高15～40cm，全株密被刚毛；茎丛生；叶倒披针形至矩圆状披针形，宽3～8mm；聚伞花序顶生，2～4分枝；花冠蓝色，辐状，5裂；小坚果背部密被瘤状凸起和刚毛，边缘具锚状刺。花期7～8月。
【分布】华北。
【生境】山地草原中的石质山坡、灌丛、石缝。
【用途】饲用。

Northern alpine forget-me-not
Borage family
Perennial herb 15～40 cm tall, densely setose throughout; stems tufted; leaves oblanceolate to oblong-lanceolate, 3～8 mm wide; cymes terminal, with 2～4 branches; corolla blue, rotate, 5-lobed; nutlets densely tuberculate and setose on back, margins with uncinate bristles. Flowering July to August.
Distribution: N China
Habitat: Rocky slopes, thickets and rock crevices in montane steppe
Use: Forage

鹤虱
Lappula myosotis Moench
紫草科
Boraginaceae

【特征】一年生或二年生草本，高20～60cm，全株密被刚毛；茎多分枝；基生叶矩圆状匙形，茎生叶披针形或条形；花冠浅蓝色，漏斗状至钟状，5裂；小坚果卵状，长3～3.5mm，具小瘤状凸起，棱缘具2行锚状刺。花期6～7月。
【分布】华北、西北；蒙古、西伯利亚、中亚、巴基斯坦、阿富汗、欧洲、北美洲。
【生境】山地草甸、河谷草甸、路旁。
【用途】药用。

Mouse-ear stickseed
Borage family
Annual or biennial herb 20～60 cm tall, densely setose throughout; stems much branched; basal leaves oblong-spatulate, cauline leaves lanceolate or linear; corolla pale blue, funnelform to campanulate, 5-lobed; nutlets ovoid, 3～3.5 mm long, tuberculate, bordered by 2 rows of uncinate bristles. Flowering June to July.
Distribution: N and NW China; Mongolia, Siberia, Central Asia, Pakistan, Afghanistan, Europe, North America
Habitat: Montane meadows, valley meadows and roadsides
Use: Medicine

拍摄人：赵利清 Photo by Zhao Liqing

滨紫草
Mertensia davurica (Sims) G. Don
紫草科
Boraginaceae

【特征】多年生草本，高20~50cm；茎直立；基生叶莲座状，密集，茎生叶近直立，披针形至条状披针形；蝎尾状聚伞花序顶生，花少数；花序轴、花梗及花萼密被伏毛；花冠蓝色，5浅裂，花冠筒部长约2cm。花期6~7月。
【分布】东北，内蒙古中东部、河北北部；蒙古、西伯利亚。
【生境】山地草甸和林缘。
【用途】饲用；观赏。

Dahurian bluebell
Borage family
Perennial herb 20~50 cm tall; stems erect; basal leaves rosulate, dense, cauline leaves subvertical, lanceolate to linear-lanceolate; scorpioid cymes terminal, with few flowers; rachis, pedicel and calyx densely appressed-hairy; corolla blue, 5-lobed, tube about 2 cm long. Flowering June to July.
Distribution: NE China, CE Inner Mongolia and N Hebei; Mongolia, Siberia
Habitat: Montane meadows and forest margins
Use: Forage; ornamental

水棘针
Amethystea caerulea L.
唇形科
Lamiaceae (Labiatae)

【特征】一年生草本，高15～40cm；茎多分枝；叶3全裂，稀5裂或不裂，边缘具锯齿，两面被柔毛；聚伞花序组成圆锥花序；花冠蓝色或蓝紫色，二唇形，上唇2裂，下唇3裂。花期7～8月。
【分布】几遍全国；蒙古、朝鲜、日本、俄罗斯、中亚、伊朗。
【生境】河岸、田边、溪旁、路边。
【用途】药用；提取芳香油。

Skyblue amethystea
Mint family
Annual herb 15～40 cm tall; stems much branched; leaves 3-divided, rarely 5-divided or undivided, margins serrate, pubescent both sides; cymes grouped in a panicle; bilabiate corolla blue or blue-purple, upper lip 2-lobed, lower lip 3-lobed. Flowering July to August.
Distribution: Almost throughout China; Mongolia, Korea, Japan, Russia, Central Asia, Iran
Habitat: River banks, farmland sides, streamsides and roadsides
Use: Medicine; extracting essential oil

毛建草 (岩青兰)
Dracocephalum rupestre Hance
唇形科
Lamiaceae (Labiatae)

【特征】多年生草本，高15～30cm；茎四棱，斜升，不分枝，被倒向短柔毛；叶三角状卵形，基部心形，边缘具圆齿；轮伞花序密集呈头状；花冠紫蓝色，二唇形，外面被短柔毛，下唇中裂片小。花期7～9月。
【分布】华北，辽宁、青海。
【生境】山地草甸、疏林、草原。
【用途】药用；观赏；茶代用品。

Cliffrock dragonhead
Mint family
Perennial herb 15～30 cm tall; stems quadrangular, ascending, unbranched, retrorsely pubescent; leaves triangular-ovate, base cordate, margins crenate; verticillasters dense to capitate; bilabiate corolla purple-blue, pubescent outside, the median lobe of lower lip smaller. Flowering July to September.
Distribution: N China, Liaoning and Qinghai
Habitat: Montane meadows, open woodlands and steppe
Use: Medicine; ornamental; substitute for tea

扭藿香
Lophanthus chinensis Benth.
唇形科
Lamiaceae (Labiatae)

【特征】多年生草本，高30～55cm；茎四棱，分枝，被柔毛和腺点；叶宽卵形至三角状卵形，边缘具圆齿，两面密被短柔毛；聚伞花序腋生，具花3～7；花冠蓝色，二唇形，冠筒扭转90°～180°，上唇转至下面，3裂，下唇转至上面，2裂。花期8～9月。
【分布】内蒙古、新疆；中亚。
【生境】山地阴坡。
【用途】观赏；提取芳香油。

Chinese gianthyssop
Mint family
Perennial herb 30～55 cm tall; stems quadrangular, branched, pubescent and glandular; leaves broadly ovate to triangular-ovate, crenate, densely pubescent; cymes axillary, with 3～7 flowers; bilabiate corolla blue, tube twisted at 90°～180°, upper lip becoming lower, 3-lobed, lower lip becoming upper, 2-lobed. Flowering August to September.
Distribution: Inner Mongolia and Xinjiang; Central Asia
Habitat: Mountain shady slopes
Use: Ornamental; extracting essential oil

拍摄人：赵利清 Photos by Zhao Liqing

大花荆芥
Nepeta sibirica L.
唇形科
Lamiaceae (Labiatae)

【特征】多年生草本，高20～70cm；茎四棱，直立或斜升；叶披针形至三角状披针形，边缘具齿，下面密被黄色腺点和微柔毛；轮伞花序顶生；花冠蓝紫色至淡紫色，外面被柔毛和腺点，二唇形，上唇2裂，下唇3裂。花期8～9月。
【分布】内蒙古西部、宁夏、甘肃北部、青海；蒙古、西伯利亚、中亚。
【生境】山坡、沟谷。
【用途】提取芳香油。

Siberian catnip
Mint family
Perennial herb 20～70 cm tall; stems quadrangular, erect or ascending; leaves lanceolate to deltoid-lanceolate, toothed, densely yellow-glandular and puberulent beneath; verticillasters terminal; bilabiate corolla blue-purple to pale purple, villous and glandular outside, upper lip 2-lobed, lower lip 3-lobed. Flowering August to September.
Distribution: W Inner Mongolia, Ningxia, N Gansu and Qinghai; Mongolia, Siberia, Central Asia
Habitat: Slopes and ravines
Use: Extracting essential oil

甘肃黄芩
Scutellaria rehderiana Diels
(阿拉善黄芩 *Scutellaria alaschanica* Tschern.)
唇形科
Lamiaceae (Labiatae)

【特征】多年生草本，高12～35cm；茎弧曲，直立，沿棱被倒向短毛；叶片卵形至披针形，全缘或中部以下边缘有浅齿；总状花序顶生；花冠粉红色、淡紫色至紫蓝色，二唇形，外面被腺毛，上唇盔状，先端微缺，下唇中裂片三角状卵形，先端微缺。花期5～8月。

【分布】内蒙古西部、山西、陕西、甘肃。

【生境】海拔1 300～3 200m的山地阳坡、岩石缝。

【用途】饲用；药用。

Gansu skullcap
Mint family
Perennial herb 12～35 cm tall; stems arcuate, erect, retrorsely pubescent on acies; leaf blades ovate to lanceolate, entire or shallowly toothed in lower half; racemes terminal; bilabiate corolla pink, pale purple to purple-blue, glandular-hairy outside, upper lip galeate with retuse apex, the median lobe of lower lip triangular-ovate with retuse apex. Flowering May to August.
Distribution: W Inner Mongolia, Shanxi, Shaanxi and Gansu
Habitat: Mountain sunny slopes and rock crevices at 1 300～3 200 m
Use: Forage; medicine

拍摄人：赵利清 Photos by Zhao Liqing

厚叶沙参
Adenophora gmelinii var. *pachyphylla* (Kitag.) Y. Z. Zhao
桔梗科
Campanulaceae

【特征】多年生草本，高40～60cm，有白色乳汁；茎直立，单一或数条；茎生叶倒披针形至倒卵状披针形，质厚，中上部边缘具齿；花序总状或花单生，花下垂；花冠蓝紫色，宽钟状，5浅裂；花柱短于花冠。花期7～8月。
【分布】东北，内蒙古、河北北部。
【生境】山地林缘、草甸。
【用途】饲用。

Thickleaf ladybells
Bellflower family
Perennial herb 40～60 cm tall, with white milky juice; stems erect, single or several; cauline leaves oblanceolate to obovate-lanceolate, thick, margin toothed in upper half; inflorescence racemose, or flower solitary, nodding; corolla blue-purple, broadly campanulate, 5-lobed; style shorter than the corolla. Flowering July to August.
Distribution: NE China, Inner Mongolia and N Hebei
Habitat: Montane forest edges and meadows
Use: Forage

草原沙参
Adenophora pratensis Y. Z. Zhao
桔梗科
Campanulaceae

【特征】多年生草本，高50～70cm，有白色乳汁；茎直立，单一；茎生叶互生，狭披针形或披针形，全缘或具疏齿；圆锥花序分枝，花下垂；花萼裂片钻状三角形；花冠蓝紫色，钟状坛形，5浅裂，裂片下部略收缢；花柱超出花冠约1/4。花期7～8月。
【分布】内蒙古（锡林郭勒）。
【生境】草原区草甸。
【用途】饲用。

Meadow ladybells
Bellflower family
Perennial herb 50～70 cm tall, with white milky juice; stems erect, single; cauline leaves alternate, narrowly lanceolate or lanceolate, entire or spaced-toothed; panicles branched, flowers nodding; sepals subutale-deltoid; corolla blue-purple, campanulate-urceolate, 5-lobed, slightly constricted below lobes; style exserted about 1/4 the corolla length. Flowering July to August.
Distribution: Inner Mongolia (Xilinguole)
Habitat: Meadows in steppe areas
Use: Forage

长柱沙参
Adenophora stenanthina (Ledeb.) Kitag.
桔梗科
Campanulaceae

【特征】多年生草本，高30～80cm，有白色乳汁；茎直立，常丛生；茎生叶互生，条形，宽2～4mm，全缘；圆锥花序顶生，多分枝，花下垂；花萼裂片钻形；花冠蓝紫色，筒状坛形，5浅裂，裂片下部略收缩；花柱超出花冠0.5～1倍。花期7～9月。
【分布】东北、华北、陕西、宁夏、甘肃、青海；蒙古、俄罗斯（西伯利亚、远东）。
【生境】山地、丘陵、沟谷、草原、灌丛、草甸、沙丘。
【用途】饲用；药用。

Narrowflower ladybells
Bellflower family
Perennial herb 30～80 cm tall, with white milky juice; stems erect, usually tufted; cauline leaves alternate, linear, 2～4 mm wide, entire; panicles terminal, much branched, flowers nodding; sepals subutale; corolla blue-purple, tubular-urceolate, 5-lobed, slightly constricted below lobes; style exserted about 0.5～1 times the corolla length. Flowering July to September.
Distribution: NE and N China, Shaanxi, Ningxia, Gansu and Qinghai; Mongolia, Russia (Siberia and Far East)
Habitat: Mountains, hills and ravines, steppe, thickets, meadows and dunes
Use: Forage; medicine

锡林沙参
Adenophora stenanthina var. *angusti-lancifolia* Y. Z. Zhao
桔梗科
Campanulaceae

【特征】与长柱沙参(*Adenophora stenanthina*)的区别为：茎生叶狭披针形或条状披针形，宽5～10mm，边缘具疏齿。
【分布】内蒙古。
【生境】沙丘间草地、山地草原。
【用途】饲用。

Xilingol ladybells
Bellflower family
Difference to *Adenophora stenanthina*: Cauline leaves narrowly lanceolate or linear-lanceolate, 5～10 mm wide, margins spaced-serrate.
Distribution: Inner Mongolia
Habitat: Meadows among dunes, montane steppe
Use: Forage

皱叶沙参
Adenophora stenanthina var. *crispata* (Turcz. ex Korsh.) Y. Z. Zhao
桔梗科
Campanulaceae

【特征】与长柱沙参(*Adenophora stenanthina*)的区别为：茎生叶披针形至卵形，宽5~15mm，边缘具波状齿。
【分布】东北、华北、陕西、宁夏。
【生境】山坡草地、疏林、沟谷、撂荒地。
【用途】饲用。

Curly ladybells
Bellflower family
Difference to *Adenophora stenanthina*: Cauline leaves lanceolate to ovate, 5~15 mm wide, margins undulate-serrate.
Distribution: NE and N China, Shaanxi and Ningxia
Habitat: Sloping grasslands, open woodlands, ravines and abandoned lands
Use: Forage

拍摄人：赵 凡 Photos by Zhao Fan

轮叶沙参
Adenophora tetraphylla (Thunb.) Fisch.
桔梗科
Campanulaceae

【特征】多年生草本，高50~90cm，有白色乳汁；茎直立，单一，不分枝；茎生叶3~6枚轮生，倒卵形至条状披针形，边缘有锯齿；花序分枝轮生，花下垂；花冠蓝色或蓝紫色，狭筒状钟形，5浅裂，裂片下部微缢缩；花柱超出花冠约1倍。花期7~8月。

【分布】东北、华北、华东、华中、华南、陕西、四川、贵州；朝鲜、日本、俄罗斯（西伯利亚、远东）、越南。

【生境】山地林缘、灌丛、草甸。

【用途】饲用；药用。

Fourleaf ladybells
Bellflower family
Perennial herb 50~90 cm tall, with white milky juice; stems erect, single, unbranched; cauline leaves in whorls of 3~6, obovate to linear-lanceolate, serrate; inflorescence with whorled branches, flowers nodding; corolla blue or blue-purple, narrowly tubular-campanulate, 5-lobed, slightly constricted below lobes; style exserted about once the corolla length. Flowering July to August.
Distribution: NE, N, E, C and S China, Shaanxi, Sichuan and Guizhou; Korea, Japan, Russia (Siberia and Far East), Viet Nam
Habitat: Montane forest margins, thickets and meadows
Use: Forage; medicine

拍摄人：赵利清 Photos by Zhao Liqing

锯齿沙参
Adenophora tricuspidata (Fisch. ex Roem. et Schult.) A. DC.
桔梗科
Campanulaceae

【特征】多年生草本，高30～60cm，有白色乳汁；茎直立，单一，不分枝；茎生叶互生，卵状披针形至条状披针形，边缘有锯齿；圆锥花序分枝短，花下垂；花萼裂片卵状三角形，下部宽而重叠，边缘有锯齿；花冠蓝色或蓝紫色，宽钟状，5浅裂；花柱短于花冠。花期7～8月。
【分布】黑龙江、内蒙古；俄罗斯（西伯利亚、远东）。
【生境】山地草甸、林缘草甸。
【用途】饲用。

Sawtooth ladybells
Bellflower family
Perennial herb 30～60 cm tall, with white milky juice; stems erect, single, unbranched; cauline leaves alternate, ovate-lanceolate to linear-lanceolate, serrate; panicles with short branches, flowers nodding; sepals ovate-triangular, base wide and overlapping, margins serrate; corolla blue or blue-purple, broadly campanulate, 5-lobed; style shorter than the corolla. Flowering July to August.
Distribution: Heilongjiang and Inner Mongolia; Russia (Siberia and Far East)
Habitat: Meadows in mountains and along forest edges
Use: Forage

菊苣
Cichorium intybus L.
菊科
Asteraceae (Compositae)

【特征】多年生草本，高30～150cm，含白色乳汁；茎直立，有细棱；基生叶莲座状，具齿或羽裂，茎生叶全缘，向上渐小；头状花序1～3，生于叶腋或单生于花序梗端；总苞圆柱形，总苞片近革质；舌状小花蓝色，先端截形，具5齿。花期6～8月。

【分布】中国北方大部分省区；亚洲、欧洲、非洲。

【生境】山坡、田野、田间、河边。

【用途】饲用；药用。

Common chicory (Blue daisy, Blue-sailors)
Aster family
Perennial herb 30～150 cm tall, with white milky juice; stems erect, thinly ribbed; basal leaves rosulate, toothed or pinnatifid, cauline leaves entire, gradually reduced upwards; heads 1～3 in leaf axils or solitary on top of peduncle; involucre cylindric, involucral bracts subleathery; ligulate florets blue, apex truncate, 5-toothed. Flowering June to August.
Distribution: Most regions of northern China; Asia, Europe, Africa
Habitat: Slopes, fields, farmlands and riversides
Use: Forage; medicine

拍摄人：赵 凡 Photos by Zhao Fan

雾灵韭
Allium stenodon Nakai et Kitag.
百合科
Liliaceae

【特征】多年生草本；鳞茎常簇生，圆柱状；叶条形，扁平；花葶高20～50cm；总苞单侧开裂，具短喙；伞形花序半球状；花被片6，蓝色至紫蓝色；花丝长于花被1～1.5倍。花期7～9月。
【分布】华北。
【生境】海拔1 550～3 000m的山地林缘和草甸。
【用途】饲用。

Wuling onion
Lily family
Perennial herb; bulbs usually clustered, cylindric; leaves linear, flat; scapes 20～50 cm tall; spathe 1-valved, short-beaked; umbel hemispheric; tepals 6, blue to purple-blue; filaments 1～1.5 times longer than the tepals. Flowering July to September.
Distribution: N China
Habitat: Montane forest margins and meadows at 1 550～3 000 m
Use: Forage

天山鸢尾
Iris loczyi Kanitz
鸢尾科
Iridaceae

【特征】多年生草本，高25～40cm；基生叶狭条形，宽1.5～3mm，平展；花葶长约15cm；花1～2枚顶生；花被片6，淡蓝色或蓝紫色，具褐色脉纹；蒴果顶端具喙。花期5～6月。
【分布】西北，内蒙古西部；中亚。
【生境】山地草原、石质山坡。
【用途】饲用；观赏。

Tianshan iris
Iris family
Perennial herb 25～40 cm tall; basal leaves narrowly linear, 1.5～3 mm wide, flat; scapes about 15 cm long; flowers 1 or 2, terminal; tepals 6, pale blue to bluish-purple, with brown streaks; capsules beaked at apex. Flowering May to June.
Distribution: NW China, W Inner Mongolia; Central Asia
Habitat: Montane steppe, rocky slopes
Use: Forage; ornamental

拍摄人：赵利清 Photo by Zhao Liqing

裸果木
Gymnocarpos przewalskii Bunge ex Maxim.
石竹科
Caryophyllaceae

【特征】半灌木，高20～80cm；分枝多而曲折；叶稍肉质，条状钻形；聚伞花序腋生；苞片膜质透明；萼片5，条形或倒披针形，红紫色，边缘膜质，先端具尖头；无花瓣；雄蕊10，外轮5雄蕊无花药。花期5～6月。
【分布】西北，内蒙古西部；蒙古。
【生境】荒漠地带的石质山坡、戈壁滩、干河床。
【用途】饲用；固沙。

Przewalsky gymnocarpos
Pink family
Subshrub 20～80 cm tall; branches many and twisty; leaves somewhat fleshy, linear-subutale; cymes axillary; bracts membranous, hyaline; sepals 5, linear or oblanceolate, reddish-purple, margins membranous, apex mucronulate; petals absent; stamens 10, the outer 5 without anther. Flowering May to June.
Distribution: NW China, W Inner Mongolia; Mongolia
Habitat: Rocky slopes, gobi and dry riverbeds in desert zone
Use: Forage; fixing dunes

紫苞风毛菊 (紫苞雪莲)
Saussurea iodostegia Hance
菊科
Asteraceae (Compositae)

【特征】多年生草本，高30～50cm；茎直立，单生，带紫色，被长柔毛；叶条状披针形至宽披针形，边缘具疏细齿，最上部叶苞叶状，膜质，紫色；头状花序4～7个，密集成伞房状；总苞片近革质，紫色，被长柔毛；管状小花紫色。花期7～8月。
【分布】东北、华北、陕西、宁夏、甘肃、四川。
【生境】山地草甸、林缘、沼泽。
【用途】饲用；观赏。

Purple saw-wort
Aster family
Perennial herb 30～50 cm tall; stems erect, single, purplish, pilose; leaves linear-lanceolate to broadly lanceolate, spaced-serrulate, uppermost leaf similar bracteal leaf, membranous, purple; heads 4～7, in a corymbiform cluster; involucral bracts subleathery, purple, pilose; tubular florets purple. Flowering July to August.
Distribution: NE and N China, Shaanxi, Ningxia, Gansu and Sichuan
Habitat: Montane meadows, forest margins and swamps
Use: Forage; ornamental

拍摄人：赵 凡 Photos by Zhao Fan

拍摄人：赵利清 Photo by Zhao Liqing

山牛蒡
Synurus deltoides (Aiton) Nakai
菊科
Asteraceae (Compositae)

【特征】多年生草本，高50～150cm；茎直立，单一，粗壮；下部叶卵形至三角形，长达20cm，边缘具缺刻状齿或羽状浅裂，具短刺，下面密被毡毛，上部叶渐小；头状花序单生茎顶，下垂；总苞被蛛丝状毛，总苞片硬而狭，顶端长刺尖状，深紫色；管状小花深紫色。花期8～9月。
【分布】东北、华北、华中；蒙古、朝鲜、日本、俄罗斯（西伯利亚、远东）。
【生境】林缘、灌丛、草甸。
【用途】药用。

Deltoid mountain thistle
Aster family
Perennial herb 50～150 cm tall; stems erect, single, stout; lower leaves ovate to triangular, to 20 cm long, margins incised-toothed or pinnatilobate, short-spinose, densely manicate below, upper leaves reduced; heads solitary and terminal, nodding; involucre arachnoid-hairy, involucral bracts rigid and narrow, apically long-pointed and dark purple; tubular florets dark purple. Flowering August to September.
Distribution: NE, N and C China; Mongolia, Korea, Japan, Russia (Siberia and Far East)
Habitat: Forest margins, thickets and meadows
Use: Medicine

藜芦
Veratrum nigrum L.
百合科
Liliaceae
【特征】多年生草本,高60～100cm;茎粗壮,直立;叶椭圆形至卵状披针形,长20～25cm,宽5～10cm;圆锥花序,花密生;花被片6,黑紫色。花期7～8月。
【分布】东北、华北、华中,四川、贵州;亚洲北部、欧洲中部。
【生境】林缘、草甸。
【用途】药用;观赏。

Black falsehellebore
Lily family
Perennial herb 60～100 cm tall; stems stout, erect; leaves elliptic to ovate-lanceolate, 20～25 cm long, 5～10 cm wide; panicles with dense flowers; tepals 6, black-purple. Flowering July to August.
Distribution: NE, N and C China, Sichuan and Guizhou; N Asia, C Europe
Habitat: Forest margins and meadows
Use: Medicine; ornamental

拍摄人：赵 凡 Photos by Zhao Fan

尖嘴苔草
Carex leiorhyncha C. A. Mey.
莎草科
Cyperaceae

【特征】多年生草本；秆高15～60cm，丛生，三棱形；叶片稍硬，扁平，两面密生锈色斑点；穗状花序圆柱状；苞片刚毛状，长于小穗；小穗多数；果囊膜质，上部具紫红色小点，边缘增厚。花期6～7月。
【分布】东北、华北、陕西、甘肃；朝鲜、俄罗斯（西伯利亚、远东）。
【生境】山地林缘草甸、溪边沼泽草甸。
【用途】饲用。

Beaked sedge
Sedge family
Perennial herb; culms 15～60 cm tall, tufted, trigonous; leaf blades slightly rigid, flat, with densely rubiginous spots; spicate inflorescence cylindric; bracts setiform, longer than spikes; spikes numerous; perigynia membranous, purple-red spotted above, margins thickened. Flowering June to July.
Distribution: NE and N China, Shaanxi and Gansu; Korea, Russia (Siberia and Far East)
Habitat: Meadows along montane forest margins, swamp-meadows along streamsides
Use: Forage

脚苔草（日荫菅、柄苔草）
Carex pediformis C. A. Mey
莎草科
Cyperaceae

【特征】多年生草本；秆高20～40cm，密丛生，纤细，钝三棱形；叶片扁平或稍对折；苞片佛焰苞状；小穗3～4枚，顶生小穗雄性，棍棒状或披针状，侧生小穗雌性，矩圆状条形；果囊中部以上密被短毛。花期5～6月。

【分布】东北、华北，陕西、甘肃、新疆；蒙古、朝鲜、俄罗斯（西伯利亚、远东）、中亚。

【生境】山地和丘陵，湿润沙地、草原、林下及林缘。

【用途】饲用。

Pediform sedge
Sedge family
Perennial herb; culms 20～40 cm tall, strongly caespitose, slender, obtusely trigonous; leaf blades flat or slightly folded; bracts spathiform; spikes 3～4, the terminal one staminate, clavate or lanceolate, the lateral ones pistillate, oblong-linear; perigynia densely short-hairy in upper half. Flowering May to June.
Distribution: NE and N China, Shaanxi, Gansu and Xinjiang; Mongolia, Korea, Russia (Siberia and Far East), Central Asia
Habitat: Mountains and hills, moist sands, steppe, forests and forest margins
Use: Forage

砾苔草
Carex stenophylloides V. I. Krecz.
莎草科
Cyperaceae

【特征】多年生草本，具根茎；秆高5～20cm，丛生，钝三棱形；叶片近扁平或内卷；苞片鳞片状；小穗3～7枚，雄雌顺序，卵形；果囊革质，具脉。花期4～6月。
【分布】东北、华北、新疆、青海、西藏；蒙古、中亚、阿富汗、伊朗、伊拉克。
【生境】草原及荒漠草原地带的沙质和砾质地、盐化草甸。
【用途】饲用。

Narrowleaf sedge
Sedge family
Perennial herb with rhizomes; culms 5～20 cm tall, tufted, obtusely trigonous; leaf blades nearly flat or involute; bracts scale-like; spikes 3～7, androgynous, ovoid; perigynia leathery, venose. Flowering April to June.
Distribution: NE and N China, Xinjiang, Qinghai and Tibet; Mongolia, Central Asia, Afghanistan, Iran, Iraq
Habitat: Sandy and gravelly sites and saline meadows in steppe and desert-steppe zones
Use: Forage

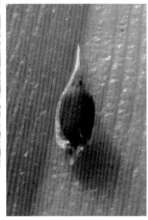

拍摄人：赵利清 Photos by Zhao Liqing

矮生嵩草
Kobresia humilis (C. A. Mey ex Trautv.) Serg.
莎草科
Cyperaceae

【特征】多年生草本，具根茎；秆高3～16cm，丛生，钝三棱形；叶片近扁平，基部对折；花序穗状，具小穗4～10余枚，顶生者雄性，侧生者雄雌顺序；先出叶矩圆形或长卵形。花期6～7月。
【分布】华北、西北、西南；蒙古、中亚。
【生境】海拔2 000～5 000m亚高山至高山带的林缘、灌丛和草甸。
【用途】饲用。

Dwarf bog sedge
Sedge family
Perennial herb with rhizomes; culms 3～16 cm tall, tufted, obtusely trigonous; leaf blades nearly flat, basally folded; inflorescence spicate with spikelets 4 to more than 10, the terminal one staminate, lateral ones androgynous; prophylls oblong or long-ovate. Flowering June to July.
Distribution: N, NW and SW China; Mongolia, Central Asia
Habitat: Forest edges, thickets and meadows in subalpine to alpine belts at 2 000～5 000 m
Use: Forage

扁秆蔍草
Scirpus planiculmis F. Schmidt
莎草科
Cyperaceae

【特征】多年生草本；根茎顶端具块茎；秆高15～85cm，单一，三棱形；叶扁平；苞片1～3，叶状，比花序长1至数倍；聚伞花序头状，有时具1～2枚辐射枝；小穗具多花；小坚果两面微凹。花期6～7月。

【分布】东北、华北、西北、华东、华中；蒙古、朝鲜、日本、俄罗斯（西伯利亚、远东）。

【生境】河边盐化草甸和沼泽。

【用途】饲用；造纸；编织；块茎药用。

Flatstem bulrush
Sedge family
Perennial herb; rhizomes with tuber at apex; culms 15～85 cm tall, single, trigonous; leaves flat; bracts 1～3, leaf-like, once to several times longer than inflorescence; cyme capitate, sometimes with rays 1 to 2; spikelets with many flowers; nutlets retuse both sides. Flowering June to July.
Distribution: NE, N, NW, E and C China; Mongolia, Korea, Japan, Russia (Siberia and Far East)
Habitat: Saline meadows and swamps along riversides
Use: Forage; papermaking; weaving; tubers for medicine

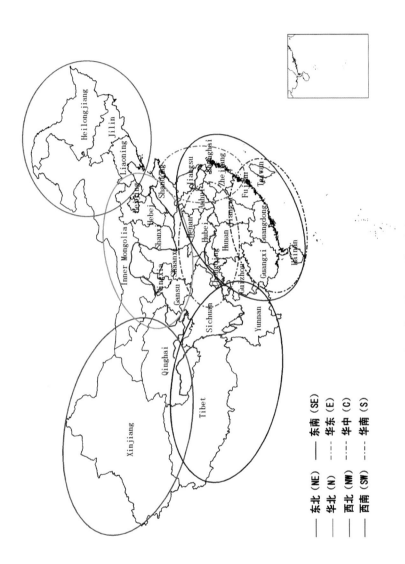

附录：中国地理区域示意图
Appendix: Geographical regions of China

参考文献 References

陈山. 中国草地饲用植物资源. 沈阳: 辽宁民族出版社, 1994
Chen Shan. Resources of Forage Plants in Chinese Grassland. Shenyang: Liaoning National Press, 1994

陈山、哈斯巴根. 蒙古高原民族植物学研究. 呼和浩特：内蒙古教育出版社，2002
Chen Shan, Khasbagan. Ethnobotanical Studies in the Mongolian Plateau. Hohhot: Inner Mongolia Education Press, 2002

耿以礼.中国主要植物图说—禾本科. 北京：科学出版社, 1959
Keng Yi-li. Flora illustralis Plantarum Primarum Sinicarum. Gramineae. Beijing: Science Press, 1959

内蒙古植物志编辑委员会. 内蒙古植物志(第二版), 第一至五卷. 呼和浩特: 内蒙古人民出版社, 1989～1994
Editorial Committee of Flora of Inner Mongolia. Flora of Inner Mongolia (the second edition), Vol. 1～5. Hohhot: Inner Mongolia Popular Press, 1989～1994

汪松、解焱.中国物种红色名录，第一卷，红色名录. 北京: 高等教育出版社, 2004
Wang Sung and Xie Yan. China Species Red List, Vol. 1 Red List. Beijing: Higher Education, 2004

新疆植物志编辑委员会.新疆植物志, 第一、二、四至六卷. 乌鲁木齐：新疆科技卫生出版社, 1992～1996
Editorial Committee of Flora of Xinjiang. Flora of Xinjiang, Vol.1, 2, 4～6. Urumqi: Xinjiang Science, Technology and Hygiene Publishing House, 1992～1996

中国科学院内蒙古宁夏综合考察队. 内蒙古植被. 北京：科学出版社, 1985
Team of the Comprehensive Scientific Expedition to Inner Mongolia-Ningxia, Chinese Academy of Sciences. Vegetation of Inner Mongolia. Beijing: Science Press, 1985

中国科学院西北高原生物研究所青海植物志编辑委员会.青海植物志, 第一至四卷. 西宁: 青海人民出版社, 1996～1999
Editorial Committee of Flora of Qinghai, Northwest Institute of Alpine Biology, Chinese Academy of Sciences. Flora of Qinghai, Vol.1～4. Xining: Qinghai Popular Press, 1996～1999

中国科学院中国植物志编辑委员会. 中国植物志(已出版各卷册). 北京：科学出版社
Editorial Committee of Flora of China, Chinese Academy of Sciences. Flora of China (the volumes published). Beijing: Science Press

Great Plains Flora Association. Flora of the Great Plains. Lawrence: University Press of Kansas, 1986

Harris J.G. and Harris M. W. (王宇飞，赵良成，冯广平和李承森 等译). 图解植物学词典. 北京：科学出版社, 2001
Harris J.G. and Harris M. W. Plant Identification Terminology: An Illustrated Glossary (Translated by: Wang Yufei, Zhao Liangcheng, Feng Guangping, Li Chengsen, etc.). Beijing: Science Press, 2001

Heath M. E., Barnes R. F. and Metcalfe D. S.. Forages, the Science of Grassland Agriculture (the fourth edition). Ames: Iowa State University Press, 1985

Tutin T. G. Et al. Flora Europaea. Cambridge: Cambridge University Press. 1964~1994, 1980

Wu Zhengyi and Petes H. Raven. Flora of China (English edition). Vol. 22. Poaceae. Beijing: Science Press, St. Louis: Missouri Botanical Garden Press, 2006

http://flower.onego.ru/, October 2010
http://plants.usda.gov, October 2010
http://www.agroatlas.ru/en/content/weeds, October 2010
http://www.cvh.org.cn/englishindex.asp, October 2010
http://www.efloras.org/index.aspx, October 2010
http://www.fna.org/, October 2010
http://www.hear.org/gcw/index.html, October 2010
http://www.ildis.org/, October 2010
http://www.tropicos.org, October 2010

中文名索引
Index of Chinese Names

A
阿尔泰葱/137
阿拉善点地梅/123
阿拉善鹅观草/54
阿拉善黄芩/331
阿拉善披碱草/54
矮生蒿草/349

B
白茎绢蒿/202
白蓝翠雀花/316
白莲蒿/181
白麻/224
白毛花旗杆/272
白屈菜/144
白羊草/46
白玉草/87
败酱/172
斑叶堇菜/292
斑子麻黄/9
抱草/64
抱茎苦荬菜/194
北齿缘草/324
北方拉拉藤/127
北庭花菜/211
北疆粉苞苣/188
北水苦荬/306
背扁黄耆/273
本氏针茅/76
扁秆藨草/350
扁茎黄耆/273
变异黄耆/274
滨紫草/326
柄苔草/347

C
草本威灵仙/307
草地婆罗门参/208
草原沙参/333
侧柏/4
叉枝鸦葱/200
叉子圆柏/6
长萼鸡眼草/278
长梗亚欧唐松草/22
长芒草/76
长毛荚黄耆/151
长腺小米草/304
长柱沙参/334
朝鲜蓟/228
车前/32
臭草/62
串珠芥/93
垂头葱芥/92
刺藜/18
刺蔷薇/242
粗ు老鹳草/283
粗壮黄耆/245
簇茎石竹/239
翠南报春/252

D
达乌里羊茅/59
大车前/35
大臭草/63
大花霸王/234
大花荆芥/330
大花雀儿豆/250
大花驼蹄瓣/234
大青山风铃草/130
大叶白麻/223
大叶蔷薇/242
大颖三芒草/42
大颖针禾/42
单脉大黄/212
单叶棘豆/281
单子麻黄/8
灯心草蚤缀/83
地构叶/28
地榆/229
垫状驼绒藜/16
东方泽泻/138
东陵八仙花/94
东陵绣球/94
豆型霸王/232
杜松/3
短梗箭头唐松草/23
短喙粉苞苣/187
短喙牦牛儿苗/109
短茎古当归/112
短龙背黄耆/117
短毛独活/117
短穗柽柳/219
短尾铁线莲/90
多茎委陵菜/147
多裂蒲公英/206

E
鄂尔多斯小聚/142
耳叶补血草/294

F
返顾马先蒿/254
飞蓬/311
肥披碱草/56

G
甘肃早雀豆/247
甘肃黄芩/331
赶山鞭/161
杠柳/253
高山地榆/217
高山蓍/132
戈壁霸王/233
戈壁雀儿豆/247
戈壁驼蹄瓣/233
茖葱/138
革苞菊/190
葛缕子/222
狗筋委陵草/87
狗尾草/74
狗牙根/51

管花秦艽/320
灌木青兰/300
灌木亚菊/174
光稃茅香/40
光稃香草/40
光稃早熟禾/69
鬼箭锦鸡儿/218

H
海韭菜/37
旱麦瓶草/85
河北假报春/251
贺兰女蒿/191
贺兰山葱/245
贺兰山南芥/91
贺兰山岩黄耆/231
褐沙蒿/178
鹤虱/325
黑柴胡/165
黑蒿/180
红柴胡/163
红早莲/160
红花海绵豆/250
红瑞木/121
厚叶花旗杆/270
厚叶沙参/332
花葩/227
花荵/322
华北八宝/216
华北大黄/12
华北覆盆子/101
华北马先蒿/256
华北前胡/119
华北乌头/314
还阳参/189
荒漠风毛菊/262
荒漠黄耆/244
黄海棠/160
黄花列当/170
黄花龙芽/172
黄精/139
黄耆/150
灰毛软紫草/298
灰叶铁线莲/141
火烙草/310

J
鸡腿堇菜/111
荠/155
戟叶鹅绒藤/297
假苇拂子茅/49
假泽早熟禾/68
尖齿糙苏/302
尖头小蘗/17
尖嘴苔草/346
碱鹅绒藤/209
碱茅/72
碱蓬/20
碱蛇床/115
渐尖风毛菊/261
箭报春/293
胶黄耆状棘豆/282

脚苔草/347
金色狗尾草/73
堇叶山梅花/96
茋草/44
荆条/299
菊苣/339
巨序剪股颖/39
苣荬菜/203
锯齿沙参/338
卷耳/84

K
苦参/108
苦苣菜/204
库页悬钩子/102
宽叶多序岩黄耆/153

L
蓝蓟/323
狼毒大戟/24
棱子芹/120
藜芦/345
砾苔草/348
镰荚棘豆/280
两栖蓼/81
了墩黄耆/246
裂叶蒿/184
裂叶堇菜/289
林木贼/2
林问荆/2
铃铛刺/248
刘氏大戟/25
柳叶风毛菊/266
柳叶芹/116
柳叶鼠李/29
龙胆/319
龙须菜/77
芦苇/66
绿花山芹/118
卵果黄耆/105
轮叶马先蒿/257
轮叶婆婆纳/307
轮叶沙参/337
裸果木/342
骆驼刺/230
骆驼蓬/110

M
麻叶荨麻/11
麦荒草/57
蔓茎蝇子草/86
猫儿菊/173
毛白花前胡/119
毛豆子/237
毛萼麦瓶草/86
毛返顾马先蒿/255
毛建草/328
毛连菜/199
毛马唐/53
毛脉酸模/15
毛蕊老鹳草/285
毛枝蒙古绣线菊/104

354

美蔷薇/243
蒙椴/158
蒙古蒿/179
蒙古绣线菊/103
蒙山莴苣/313
密花风毛菊/261
密毛白莲蒿/182
漠蒿/177
木藤蓼/80
木藤首乌/80
木贼麻黄/7
牧地山黧豆/154
墓头回/171

N
内蒙野丁香/258
宁夏沙参/308
牛蒡/259
牛扁/140
扭果花旗杆/271
扭藿香/329
女娄菜/214

P
平车前/33
铺散亚菊/175
蒲公英/207

Q
歧伞獐牙菜/125
千叶蓍/226
茜草/128
茄叶碱蓬/21
秦艽/318
球果堇菜/288
瞿麦/240
曲枝天门冬/78
全缘橐吾/198

R
日荫菅/347
乳苣/313

S
三花龙胆/321
三脉紫菀/309
三芒草/41
沙地柏/6
沙地繁缕/88
沙地青兰/300
沙地雀麦/48
沙蒿/177
沙茴香/166
沙棘/122
山蒿/176
山苦荬/193
山柳菊/190
山蚂蚱草/85
山牛蒡/344
山莴苣/312
芍药/89
少花顶冰花/210
少花米口袋/277
湿地黄耆/152
蓍/226
石生霸王/236
石生驼蹄瓣/236
石竹/238
双花堇菜/162
水棘针/327
水金凤/157
水麦冬/38
水枸子/97

丝路蓟/260
宿根亚麻/317
酸枣/31

T
蹄叶橐吾/196
天山鸢尾/341
天仙子/168
田葛缕子/114
条叶车前/34
铁杆蒿/181
头花丝石竹/213
头状石头花/213
突节老鹳草/284
团球火绒草/195
驼蹄瓣/
椭圆叶花锚/295
椭圆叶天芥菜/126

W
万年蒿/181
伪泥胡菜/268
委陵菜/146
蚊子草/99
问荆/1
乌拉特黄耆/245
乌腺金丝桃/161
无毛兔唇花/225
雾灵韭/340

X
西北风毛菊/264
西山委陵菜/149
锡林沙参/335
细叉梅花草/95
细穗柽柳/220
细叶鸢/133
细叶小檗/143
细叶早熟禾/67
狭苞橐吾/197
狭叶荨麻/10
下延叶古当归/113
纤细绢蒿/201
线棘豆/249
线叶蒿/183
小点地梅/124
小裂叶荆芥/303
小米草/305
小叶忍冬/129
小叶鼠李/30
楔叶菊/134
新疆方枝柏/5
新疆花葵/287
新疆芍药/241
新疆野百合/269
兴安柴胡/164
兴安女娄菜/215
星毛短舌菊/186
秀丽马先蒿/169

Y
鸭茅/52
雅布赖风毛菊/267
亚洲蒲公英/205
岩蒿/176
岩青兰/328
盐乏木/248
盐生车前/36
羊角子草/296
野黍/58
野西瓜苗/159
野亚麻/286
野燕麦/45

一叶萩/27
异燕麦/61
异叶败酱/171
异针茅/75
翼果霸王/235
翼果驼蹄瓣/235
藕草/65
阴山鸟头/315
银露梅100
蚓果芥/93
隐花草/50
樱草/252
硬阿魏/166
硬尖神香草/301
硬叶风毛菊/263
硬叶乌苏里风毛菊/263
硬质早熟禾/70
尤那托夫黄耆/106
羽毛三芒草/43
羽毛针禾/43
羽毛鬼针草/185
玉门黄耆/275
圆果黄耆/106
圆叶木蓼/79
圆柱披碱草/55
缘毛雀麦/47

Z
杂交景天/145
杂配藜/19
藏蓟牛儿苗/109
早开堇菜/291
扎股草/50
展毛黄芩/167
掌叶多裂委陵菜/148
胀果甘草/276
沼生柳叶菜/221
沼泽蒿/180
折苞风毛菊/265
针枝芸香/156
中亚细柄茅/71
皱叶沙参/336
皱叶酸模/14
珠芽蓼/82
准噶尔大戟/26
准噶尔枸子/98
紫斑风铃草/131
紫苞风毛菊/343
紫苞雪莲/343
紫苜蓿/279
紫羊茅/60
总裂叶堇菜/290
总序大黄/13
总状土木香/192

拉丁名索引
Index of Latin Names

A

Achillea alpina /132
Achillea millefolium /226
Achyrophorus ciliatus /173
Aconitum barbatum var. puberulum /140
Aconitum jeholense var. angustius /314
Aconitum yinschanicum /315
Adenophora gmelinii var. pachyphylla /332
Adenophora ningxianica /308
Adenophora pratensis /333
Adenophora stenanthina var. angusti-lancifolia /335
Adenophora stenanthina var. crispata /336
Adenophora stenanthina /334
Adenophora tetraphylla /337
Adenophora tricuspidata /338
Agrostis gigantea /39
Ajania fruticulosa /174
Ajania khartensis /175
Alhagi sparsifolia /230
Alisma orientale /138
Allium altaicum /137
Allium sacculiferum /228
Allium stenodon /340
Allium victorialis /138
Amethystea caerulea /327
Androsace alaschanica /123
Androsace gmelinii /124
Anthoxanthum glabrum /40
Apocynum hendersonii /223
Apocynum pictum /224
Arabis alaschanica /91
Arabis pendula /92
Archangelica brevicaulis /112
Archangelica decurrens /113
Arctium lappa /259
Arenaria juncea /83
Aristida adscensionis /41
Aristida grandiglumis /42
Aristida pennata /43
Arnebia fimbriata /298
Artemisia brachyloba /176
Artemisia desertorum /177
Artemisia intramongolica /178
Artemisia mongolica /179
Artemisia palustris /180
Artemisia sacrorum var. messerschmidtiana /182
Artemisia sacrorum /181
Artemisia subulata /183
Artemisia tanacetifolia /184

Arthraxon hispidus /44
Asparagus schoberioides /77
Asparagus trichophyllus /78
Aster ageratoides /309
Astragalus alaschanensis /244
Astragalus complanatus /273
Astragalus dengkouensis /244
Astragalus grubovii /105
Astragalus hoantchy /245
Astragalus junatovii /106
Astragalus lioui /246
Astragalus membranaceus /150
Astragalus monophyllus /151
Astragalus parvicarinatus /107
Astragalus pavlovii /246
Astragalus uliginosus /152
Astragalus variabilis /274
Astragalus yumenensis /275
Atraphaxis tortuosa /79
Avena fatua /45

B

Berberis poiretii /143
Berberis carolii /142
Bidens maximowicziana /185
Bothriochloa ischaemum /46
Brachanthemum pulvinatum /186
Bromus ciliatus /47
Bromus ircutensis /48
Bupleurum scorzonerifolium /163
Bupleurum sibiricum /164
Bupleurum smithii /165
Butomus umbellatus /227

C

Calamagrostis pseudophragmites /49
Campanula glomerata ssp. daqingshanica /130
Campanula punctata /131
Caragana jubata /218
Carex leiorhyncha /346
Carex pediformis /347
Carex stenophylloides /348
Carum buriaticum /114
Carum carvi /222
Cerastium arvense /84
Ceratoides compacta /16
Chelidonium majus /144
Chenopodium acuminatum /17
Chenopodium aristatum /18

356

Chenopodium hybridum /19
Chesneya grubovii /247
Chesneya macrantha /250
Chesniella ferganensis /247
Chondrilla brevirostris /187
Chondrilla lejosperma /188
Cichorium intybus /339
Cirsium arvense /260
Clematis brevicaudata /90
Clematis canescens /141
Cnidium salinum /115
Cornus alba /121
Cornus bretschneideri /122
Cortusa matthioli ssp. pekinensis /251
Cotoneaster multiflorus /97
Cotoneaster soongoricus /98
Crepis crocea /189
Crypsis aculeata /50
Cynanchum cathayense /296
Cynanchum sibiricum /297
Cynodon dactylon /51
Czernaevia laevigata /116

D

Dactylis glomerata /52
Delphinium albocoeruleum /316
Dendranthema maximowiczii /133
Dendranthema naktongense /134
Dianthus chinensis /238
Dianthus repens /239
Dianthus superbus /240
Digitaria ciliaris var. chrysoblephara /53
Dontostemon crassifolius /270
Dontostemon elegans /271
Dontostemon senilis /272
Dracocephalum fruticulosum /300
Dracocephalum psammophilum /300
Dracocephalum rupestre /328

E

Echinops przewalskii /310
Echium vulgare /323
Elymus alashanicus /54
Elymus cylindricus /55
Elymus dahuricus var. cylindricus /55
Elymus excelsus /56
Elymus tangutorum /57
Ephedra equisetina /7
Ephedra monosperma /8
Ephedra rhytidosperma /9
Epilobium palustre /221
Equisetum arvense /1
Equisetum sylvaticum /2
Erigeron acer /311
Eriochloa villosa /58
Eritrichium borealisinense /324
Erodium tibetanum /109
Euphorbia fischeriana /24

Euphorbia lioui /25
Euphorbia soongarica /26
Euphrasia hirtella /304
Euphrasia pectinata /305

F

Fallopia aubertii /80
Ferula bungeana /166
Festuca dahurica /59
Festuca rubra /60
Filipendula palmata /99
Flueggea suffruticosa /27

G

Gagea pauciflora /210
Galium boreale /127
Gentiana macrophylla /318
Gentiana scabra /319
Gentiana siphonantha /320
Gentiana triflora /321
Geranium dahuricum /283
Geranium krameri /284
Geranium platyanthum /285
Glycyrrhiza inflata /276
Gueldenstaedtia verna /277
Gymnocarpos przewalskii /342
Gypsophila capituliflora /213

H

Halenia elliptica /295
Halimodendron halodendron /248
Haplophyllum tragacanthoides /156
Hedysarum petrovii /231
Hedysarum polybotrys var. alaschanicum /153
Helictotrichon schellianum /61
Heliotropium ellipticum /126
Hemerocallis lilioasphodelus /211
Heracleum moellendorffii /117
Hibiscus trionum /159
Hieracium umbellatum /190
Hierochloe glabra /40
Hippolytia alashanensis /191
Hydrangea bretschneideri /94
Hylotelephium tatarinowii /216
Hyoscyamus niger /168
Hypericum ascyron /160
Hypericum attenuatum /161
Hypochaeris ciliata /173
Hyssopus cuspidatus /301

I

Impatiens noli-tangere /157
Inula racemosa /192
Iris loczyi /341
Ixeris chinensis /193
Ixeris sonchifolia /194

J

Juniperus pseudosabina /5
Juniperus rigida /3
Juniperus sabina /6

K

Kobresia humilis /349
Krascheninnikovia compacta /16
Kummerowia stipulacea /278

L

Lactuca sibirica /312
Lactuca tatarica /313
Lagedium sibiricum /312
Lagochilus bungei /225
Lappula myosotis /325
Lathyrus pratensis /154
Lavatera cashemiriana /287
Leontopodium conglobatum /195
Leptodermis ordosica /258
Ligularia fischeri /196
Ligularia intermedia /197
Ligularia mongolica /198
Lilium dauricum /237
Lilium martagon var. pilosiusculum /269
Limonium otolepis /294
Linum perenne /317
Linum stelleroides /286
Lonicera microphylla /129
Lophanthus chinensis /329
Lychnis brachypetala /215

M

Medicago sativa /279
Melandrium apricum /214
Melandrium brachypetalum /215
Melica scabrosa /62
Melica turczaninowiana /63
Melica virgata /64
Mertensia davurica /326
Mulgedium tataricum /313

N

Neotorularia humilis /93
Nepeta sibirica /330

O

Orobanche pycnostachya /170
Ostericum viridiflorum /118
Oxytropis falcata /280
Oxytropis filiformis /249
Oxytropis monophylla /281
Oxytropis tragacanthoides /282

P

Paeonia lactiflora /89
Paeonia sinjiangensis /241

Parnassia oreophila /95
Patrinia heterophylla /171
Patrinia scabiosifolia /172
Pedicularis resupinata var. pubescens /255
Pedicularis resupinata /254
Pedicularis tatarinowii /256
Pedicularis venusta /169
Pedicularis verticillata /257
Peganum harmala /110
Periploca sepium /253
Peucedanum harry-smithii /119
Peucedanum praeruptorum ssp. hirsutiusculum /119
Phalaris arundinacea /65
Philadelphus tenuifolius /96
Phlomis dentosa /302
Phragmites australis /66
Picris davurica /199
Plantago asiatica /32
Plantago depressa /33
Plantago lessingii /34
Plantago major /35
Plantago maritima /36
Platycladus orientalis /4
Pleurospermum uralense /120
Poa angustifolia /67
Poa pratensis ssp. angustifolia /67
Poa pseudopalustris /68
Poa psilolepis /69
Poa sphondylodes /70
Poacynum hendersonii /223
Poacynum pictum /224
Polemonium caeruleum /322
Polygonatum sibiricum /139
Polygonum amphibium /81
Polygonum aubertii /80
Polygonum viviparum /82
Potentilla chinensis /146
Potentilla glabra /100
Potentilla multicaulis /147
Potentilla multifida var. ornithopoda /148
Potentilla sischanensis /149
Primula fistulosa /293
Primula sieboldii /252
Ptilagrostis pelliotii /71
Puccinellia distans /72

R

Rhamnus erythroxylum /29
Rhamnus parvifolia /30
Rheum franzenbachii /12
Rheum racemiferum /13
Rheum uninerve /212
Roegneria alashanica /54
Rosa acicularis /242
Rosa bella /243
Rubia cordifolia /128
Rubus idaeus var. borealisinensis /101
Rubus sachalinensis /102

Rumex crispus /14
Rumex gmelinii /15

S

Sabina pseudosabina /5
Sabina vulgaris /6
Sanguisorba alpina /217
Sanguisorba officinalis /229
Saussurea acuminata /261
Saussurea deserticola /262
Saussurea firma /263
Saussurea iodostegia /343
Saussurea petrovii /264
Saussurea recurvata /265
Saussurea salicifolia /266
Saussurea yabulaiensis /267
Schizonepeta annua /303
Scirpus planiculmis /350
Scorzonera muriculata /200
Scutellaria alaschanica /331
Scutellaria orthotricha /167
Scutellaria rehderiana /331
Sedum hybridum /145
Seriphidium gracilescens /201
Seriphidium terrae-albae /202
Serratula coronata /268
Setaria glauca /73
Setaria viridis /74
Silene aprica /214
Silene jenisseensis /85
Silene repens /86
Silene venosa /87
Sonchus arvensis /203
Sonchus oleraceus /204
Sophora flavescens /108
Speranskia tuberculata /28
Spiraea mongolica var. tomentulosa /104
Spiraea mongolica /103
Spongiocarpella grubovii /250
Stellaria gypsophiloides /88
Stipa aliena /75
Stipa bungeana /76
Stipagrostis grandiglumis /42
Stipagrostis pennata /43
Suaeda glauca /20
Suaeda przewalskii /21
Swertia dichotoma /125
Swida alba /121
Swida bretschneideri /122
Synurus deltoides /344

T

Tamarix laxa /219
Tamarix leptostachya /220
Taraxacum asiaticum /205
Taraxacum dissectum /206
Taraxacum mongolicum /207
Thalictrum minus var. stipellatum /22

Thalictrum simplex var. brevipes /23
Tilia mongolica /158
Torularia humilis /93
Tragopogon pratensis /208
Tribulus terrestris /155
Triglochin maritima /37
Triglochin palustris /38
Tugarinovia mongolica /135

U

Urtica angustifolia /10
Urtica cannabina /11

V

Veratrum nigrum /345
Veronica anagallis-aquatica /306
Veronicastrum sibiricum /307
Viola acuminata /111
Viola biflora /162
Viola collina /288
Viola dissecta /289
Viola fissifolia /290
Viola prionantha /291
Viola variegata /292
Vitex negundo var. heterophylla /299

Y

Youngia stenoma /209

Z

Ziziphus jujuba var. spinosa /31
Zygophyllum fabago /232
Zygophyllum gobicum /233
Zygophyllum potaninii /234
Zygophyllum pterocarpum /235
Zygophyllum rosowii /236

拉丁名及中文名分科索引
Index of Latin and Chinese Names in Families

1 木贼科 Equisetaceae
    Equisetum arvense 问荆/1
    Equisetum sylvaticum 林问荆(林木贼)/2

2 柏科 Cupressaceae
    Juniperus rigida 杜松/3
    Platycladus orientalis 侧柏/4
    Sabina pseudosabina 新疆方枝柏
    (Juniperus pseudosabina)/5
    Sabina vulgaris 叉子圆柏
    (Juniperus sabina 沙地柏)/6

3 麻黄科 Ephedraceae
    Ephedra equisetina 木贼麻黄/7
    Ephedra monosperma 单子麻黄/8
    Ephedra rhytidosperma 斑子麻黄/9

4 荨麻科 Urticaceae
    Urtica angustifolia 狭叶荨麻/10
    Urtica cannabina 麻叶荨麻/11

5 蓼科 Polygonaceae
    Atraphaxis tortuosa 圆叶木蓼/79
    Fallopia aubertii 木藤首乌
    (Polygonum aubertii 木藤蓼)/80
    Polygonum amphibium 两栖蓼/81
    Polygonum viviparum 珠芽蓼/82
    Rheum franzenbachii 华北大黄/12
    Rheum racemiferum 总序大黄/13
    Rheum uninerve 单脉大黄/212
    Rumex crispus 皱叶酸模/14
    Rumex gmelinii 毛脉酸模/15

6 藜科 Chenopodiaceae
    Ceratoides compacta 垫状驼绒藜
    (Krascheninnikovia compacta)/16
    Chenopodium acuminatum 尖头叶藜/17
    Chenopodium aristatum 刺藜/18
    Chenopodium hybridum 杂配藜/19
    Suaeda glauca 碱蓬/20
    Suaeda przewalskii 茄叶碱蓬/21

7 石竹科 Caryophyllaceae
    Arenaria juncea 灯心草蚤缀/83
    Cerastium arvense 卷耳/84
    Dianthus chinensis 石竹/238
    Dianthus repens 簇茎石竹/239
    Dianthus superbus 瞿麦/240
    Gymnocarpos przewalskii 裸果木/342

    Gypsophila capituliflora
    头状石头花(头花丝石竹)/213
    Melandrium apricum 女娄菜 (Silene aprica)/214
    Melandrium brachypetalum 兴安女娄菜
    (Lychnis brachypetala)/215
    Silene jenisseensis 旱麦瓶草(山蚂蚱草)/85
    Silene repens 毛萼麦瓶草(蔓茎蝇子草)/86
    Silene venosa 狗筋麦瓶草(白玉草)/87
    Stellaria gypsophiloides 沙地繁缕/88

8 毛茛科 Ranunculaceae
    Aconitum barbatum var. puberulum 牛扁/140
    Aconitum jeholense var. angustius 华北乌头/314
    Aconitum yinschanicum 阴山乌头/315
    Clematis brevicaudata 短尾铁线莲/89
    Clematis canescens 灰叶铁线莲/141
    Delphinium albocoeruleum 白蓝翠雀花/316
    Thalictrum minus var. stipellatum
    长梗亚欧唐松草/22
    Thalictrum simplex var. brevipes
    短梗箭头唐松草/23

9 芍药科（牡丹科）Paeoniaceae
    Paeonia lactiflora 芍药/90
    Paeonia sinjiangensis 新疆芍药/241

10 小檗科 Berberidaceae
    Berberis caroli 鄂尔多斯小檗/142
    Berberis poiretii 细叶小檗/143

11 罂粟科 Papaveraceae
    Chelidonium majus 白屈菜/144

12 十字花科 Brassicaceae
    Arabis alaschanica 贺兰山南芥/91
    Arabis pendula 垂果南芥/92
    Dontostemon crassifolius 厚叶花旗杆/270
    Dontostemon elegans 扭果花旗杆/271
    Dontostemon senilis 白毛花旗杆/272
    Neotorularia humilis 串珠芥
    (Torularia humilis 蚓果芥)/93

13 景天科 Crassulaceae
    Hylotelephium tatarinowii 华北八宝/216
    Sedum hybridum 杂交景天/145

14 虎耳草科 Saxifragaceae
    Hydrangea bretschneideri
    东陵八仙花(东陵绣球)/94
    Parnassia oreophila 细叉梅花草/95
    Philadelphus tenuifolius 薄叶山梅花/96

15 蔷薇科 Rosaceae

    Cotoneaster multiflorus 水栒子/97
    Cotoneaster soongoricus 准噶尔栒子/98
    Filipendula palmata 蚊子草/99
    Potentilla chinensis 委陵菜/146
    Potentilla glabra 银露梅100
    Potentilla multicaulis 多茎委陵菜/147
    Potentilla multifida var. ornithopoda 掌叶多裂委陵菜/148
    Potentilla sischanensis 西山委陵菜/149
    Rosa acicularis 刺蔷薇(大叶蔷薇)/242
    Rosa bella 美蔷薇/243
    Rubus idaeus var.borealisinensis 华北覆盆子/101
    Rubus sachalinensis 库页悬钩子/102
    Sanguisorba alpina 高山地榆/217
    Sanguisorba officinalis 地榆/229
    Spiraea mongolica 蒙古绣线菊/103
    Spiraea mongolica var. tomentulosa 毛枝蒙古绣线菊/104

16 豆科 Fabaceae (Leguminosae)

    Alhagi sparsifolia 骆驼刺/230
    Astragalus alaschanensis 荒漠黄耆 (Astragalus dengkouensis)/244
    Astragalus complanatus 扁茎黄耆(背扁黄耆)/273
    Astragalus grubovii 卵果黄耆/105
    Astragalus hoantchy 粗壮黄耆(乌拉特黄耆、贺兰山黄耆)/245
    Astragalus junatovii 圆果黄耆(尤那托夫黄耆)/106
    Astragalus membranaceus 黄耆/150
    Astragalus monophyllus 长毛荚黄耆/151
    Astragalus parvicarinatus 短龙骨黄耆/107
    Astragalus pavlovii 了墩黄耆(Astragalus lioui)/246
    Astragalus uliginosus 湿地黄耆/152
    Astragalus variabilis 变异黄耆/274
    Astragalus yumenensis 玉门黄耆/275
    Caragana jubata 鬼箭锦鸡儿/218
    Chesniella ferganensis 甘肃旱雀豆 (Chesneya grubovii 戈壁雀儿豆)/247
    Glycyrrhiza inflata 胀果甘草/276
    Gueldenstaedtia verna 少花米口袋/277
    Halimodendron halodendron 盐豆木(铃铛刺)/248
    Hedysarum petrovii 贺兰山岩黄耆/231
    Hedysarum polybotrys var. alaschanicum 宽叶多序岩黄耆/153
    Kummerowia stipulacea 长萼鸡眼草/278
    Lathyrus pratensis 牧地山黧豆/154
    Medicago sativa 紫花苜蓿/279
    Oxytropis falcata 镰荚棘豆/280
    Oxytropis filiformis 线棘豆/249
    Oxytropis monophylla 单叶棘豆/281
    Oxytropis tragacanthoides 胶黄耆状棘豆/282
    Sophora flavescens 苦参/108
    Spongiocarpella grubovii 红花海绵豆 (Chesneya macrantha 大花雀儿豆)/250

17 牻牛儿苗科 Geraniaceae

    Erodium tibetanum 短喙牻牛儿苗(藏牻牛儿苗)/109
    Geranium dahuricum 粗根老鹳草/283
    Geranium krameri 突节老鹳草/284
    Geranium platyanthum 毛蕊老鹳草/285

18 亚麻科 Linaceae

    Linum perenne 宿根亚麻/317
    Linum stelleroides 野亚麻/286

19 蒺藜科 Zygophyllaceae

    Peganum harmala 骆驼蓬/110
    Tribulus terrestris 蒺藜/155

    Zygophyllum fabago 驼蹄瓣(豆型霸王)/232
    Zygophyllum gobicum 戈壁驼蹄瓣(戈壁霸王)/233
    Zygophyllum potaninii 大花驼蹄瓣(大花霸王)/234
    Zygophyllum pterocarpum 翼果驼蹄瓣(翼果霸王)/235
    Zygophyllum rosowii 石生驼蹄瓣(石生霸王)/236

20 芸香科 Rutaceae

    Haplophyllum tragacanthoides 针枝芸香/156

21 大戟科 Euphorbiaceae

    Euphorbia fischeriana 狼毒大戟/24
    Euphorbia lioui 刘氏大戟/25
    Euphorbia soongarica 准噶尔大戟/26
    Flueggea suffruticosa 一叶萩/27
    Speranskia tuberculata 地构叶/28

22 凤仙花科 Balsaminaceae

    Impatiens noli-tangere 水金凤/157

23 鼠李科 Rhamnaceae

    Rhamnus erythroxylum 柳叶鼠李/29
    Rhamnus parvifolia 小叶鼠李/30
    Ziziphus jujuba var. spinosa 酸枣/31

24 椴树科 Tiliaceae

    Tilia mongolica 蒙椴/158

25 锦葵科 Malvaceae

    Hibiscus trionum 野西瓜苗/159
    Lavatera cashemiriana 新疆花葵/287

26 金丝桃科 Hypericaceae

    (藤黄科 Clusiaceae)

    Hypericum ascyron 黄海棠(红旱莲)/160
    Hypericum attenuatum 乌腺金丝桃(赶山鞭)/161

27 柽柳科 Tamaricaceae

    Tamarix laxa 短穗柽柳/219
    Tamarix leptostachya 细穗柽柳/220

28 堇菜科 Violaceae

    Viola acuminata 鸡腿堇菜/111
    Viola biflora 双花堇菜/162
    Viola collina 球果堇菜/288
    Viola dissecta 裂叶堇菜/289
    Viola fissifolia 总裂叶堇菜/290
    Viola prionantha 早开堇菜/291
    Viola variegata 斑叶堇菜/292

29 柳叶菜科 Onagraceae

    Epilobium palustre 沼生柳叶菜/221

30 伞形科 Apiaceae (Umbelliferae)

    Archangelica brevicaulis 短茎古当归/112
    Archangelica decurrens 下延古当归/113
    Bupleurum scorzonerifolium 红柴胡/163
    Bupleurum sibiricum 兴安柴胡/164
    Bupleurum smithii 黑柴胡/165
    Carum buriaticum 田葛缕子/114
    Carum carvi 葛缕子/222
    Cnidium salinum 碱蛇床/115
    Czernaevia laevigata 柳叶芹/116
    Ferula bungeana 沙茴香(硬阿魏)/166

Heracleum moellendorffii 短毛独活/117
Ostericum viridiflorum 绿花山芹/118
Peucedanum harry-smithii 华北前胡
(Peucedanum praeruptorum ssp. hirsutiusculum
毛白花前胡)/119
Pleurospermum uralense 棱子芹/120

31 山茱萸科 Cornaceae

Swida alba 红瑞木 (Cornus alba)/121
Swida bretschneideri 沙梾
(Cornus bretschneideri)/122

32 报春花科 Primulaceae

Androsace alaschanica 阿拉善点地梅/123
Androsace gmelinii 小点地梅/124
Cortusa matthioli ssp. pekinensis
河北假报春/251
Primula fistulosa 箭报春/293
Primula sieboldii 翠南报春(樱草)/252

33 白花丹科 Plumbaginaceae

Limonium otolepis 耳叶补血草/294

34 龙胆科 Gentianaceae

Gentiana macrophylla 秦艽/318
Gentiana scabra 龙胆/319
Gentiana siphonantha 管花秦艽/320
Gentiana triflora 三花龙胆/321
Halenia elliptica 椭圆叶花锚/295
Swertia dichotoma 歧伞獐牙菜/125

35 夹竹桃科 Apocynaceae

Poacynum hendersonii 大叶白麻
(Apocynum hendersonii)/223
Poacynum pictum 白麻 (Apocynum pictum)/224

36 萝藦科 Asclepiadaceae

Cynanchum cathayense 羊角子草/296
Cynanchum sibiricum 戟叶鹅绒藤/297
Periploca sepium 杠柳/253

37 花荵科 Polemoniaceae

Polemonium caeruleum 花荵/322

38 紫草科 Boraginaceae

Arnebia fimbriata 灰毛软紫草/298
Echium vulgare 蓝蓟/323
Eritrichium borealisinense 北齿缘草/324
Heliotropium ellipticum 椭圆叶天芥菜/126
Lappula myosotis 鹤虱/325
Mertensia davurica 滨紫草/326

39 马鞭草科 Verbenaceae

Vitex negundo var. heterophylla 荆条/299

40 唇形科 Lamiaceae (Labiatae)

Amethystea caerulea 水棘针/327
Dracocephalum fruticulosum 灌木青兰
(Dracocephalum psammophilum 沙地青兰)/300
Dracocephalum rupestre 毛建草(岩青兰)/328
Hyssopus cuspidatus 硬尖神香草/301
Lagochilus bungei 无毛兔唇花/225
Lophanthus chinensis 扭藿香/329
Nepeta sibirica 大花荆芥/330
Phlomis dentosa 尖齿糙苏/302

Schizonepeta annua 小裂叶荆芥/303
Scutellaria orthotricha 展毛黄芩/167
Scutellaria rehderiana 甘肃黄芩
(Scutellaria alaschanica 阿拉善黄芩)/331

41 茄科 Solanaceae

Hyoscyamus niger 天仙子/168

42 玄参科 Scrophulariaceae

Euphrasia hirtella 长腺小米草/304
Euphrasia pectinata 小米草/305
Pedicularis resupinata 返顾马先蒿/254
Pedicularis resupinata var. pubescens
毛返顾马先蒿/255
Pedicularis tatarinowii 华北马先蒿/256
Pedicularis venusta 秀丽马先蒿/169
Pedicularis verticillata 轮叶马先蒿/257
Veronica anagallis-aquatica 北水苦荬/306
Veronicastrum sibiricum
草本威灵仙(轮叶婆婆纳)/307

43 列当科 Orobanchaceae

Orobanche pycnostachya 黄花列当/170

44 车前科 Plantaginaceae

Plantago asiatica 车前/32
Plantago depressa 平车前/33
Plantago lessingii 条叶车前/34
Plantago major 大车前/35
Plantago maritima 盐生车前/36

45 茜草科 Rubiaceae

Galium boreale 北方拉拉藤/127
Leptodermis ordosica 内蒙野丁香/258
Rubia cordifolia 茜草/128

46 忍冬科 Caprifoliaceae

Lonicera microphylla 小叶忍冬/129

47 败酱科 Valerianaceae

Patrinia heterophylla 异叶败酱(墓头回)/171
Patrinia scabiosifolia 败酱(黄花龙芽)/172

48 桔梗科 Campanulaceae

Adenophora gmelinii var. pachyphylla 厚叶沙参/332
Adenophora ningxianica 宁夏沙参/308
Adenophora pratensis 草原沙参/333
Adenophora stenanthina 长柱沙参/334
Adenophora stenanthina var. angusti-lancifolia
锡林沙参/335
Adenophora stenanthina var. crispata 皱叶沙参/336
Adenophora tetraphylla 轮叶沙参/337
Adenophora tricuspidata 锯齿沙参/338
Campanula glomerata subsp. daqingshanica
大青山风铃草/130
Campanula punctata 紫斑风铃草/131

49 菊科 Asteraceae (Compositae)

Achillea alpina 高山蓍/132
Achillea millefolium 蓍(千叶蓍)/226
Achyrophorus ciliatus 猫儿菊
(Hypochaeris ciliati)/173
Ajania fruticulosa 灌木亚菊/174
Ajania khartensis 铺散亚菊/175
Arctium lappa 牛蒡/259
Artemisia brachyloba 山蒿(岩蒿)/176

Artemisia desertorum 漠蒿(沙蒿)/177
Artemisia intramongolica 褐沙蒿/178
Artemisia mongolica 蒙古蒿/179
Artemisia palustris 黑蒿(沼泽蒿)/180
Artemisia sacrorum 白莲蒿(万年蒿、铁杆蒿)/181
Artemisia sacrorum var. messerschmidtiana
密毛白莲蒿/182
Artemisia subulata 线叶蒿/183
Artemisia tanacetifolia 裂叶蒿/184
Aster ageratoides 三脉紫菀/309
Bidens maximowicziana 羽叶鬼针草/185
Brachanthemum pulvinatum 星毛短舌菊/186
Chondrilla brevirostris 短喙粉苞苣/187
Chondrilla lejosperma 北疆粉苞苣/188
Cichorium intybus 菊苣/339
Cirsium arvense 丝路蓟/260
Crepis crocea 还阳参/189
Dendranthema maximowiczii 细叶菊/133
Dendranthema naktongense 楔叶菊/134
Echinops przewalskii 火烙草/310
Erigeron acer 飞蓬/311
Hieracium umbellatum 山柳菊/190
Hippolytia alashanensis 贺兰女蒿/191
Inula racemosa 总状土木香/192
Ixeris chinensis 山苦荬/193
Ixeris sonchifolia 抱茎苦荬菜/194
Lagedium sibiricum 山莴苣 (Lactuca sibirica)/312
Leontopodium conglobatum 团球火绒草/195
Ligularia fischeri 蹄叶橐吾/196
Ligularia intermedia 狭苞橐吾/197
Ligularia mongolica 全缘橐吾/198
Mulgedium tataricum 乳苣
(Lactuca tatarica 蒙山莴苣)/313
Picris davurica 毛连菜/199
Saussurea acuminata 密花风毛菊(渐尖风毛菊)/261
Saussurea deserticola 荒漠风毛菊/262
Saussurea firma 硬叶风毛菊(硬叶乌苏里风毛菊)/263
Saussurea iodostegia 紫苞风毛菊(紫苞雪莲)/343
Saussurea petrovii 西北风毛菊/264
Saussurea recurvata 折苞风毛菊/265
Saussurea salicifolia 柳叶风毛菊/266
Saussurea yabulaiensis 雅布赖风毛菊/267
Scorzonera muriculata 叉枝鸦葱/200
Seriphidium gracilescens 纤细绢蒿/201
Seriphidium terrae-albae 白茎绢蒿/202
Serratula coronata 伪泥胡菜/268
Sonchus arvensis 苣荬菜/203
Sonchus oleraceus 苦苣菜/204
Synurus deltoides 山牛蒡/344
Taraxacum asiaticum 亚洲蒲公英/205
Taraxacum dissectum 多裂蒲公英/206
Taraxacum mongolicum 蒲公英/207
Tragopogon pratensis 草地婆罗门参/208
Tugarinovia mongolica 革苞菊/135
Youngia stenoma 碱黄鹌菜/209

50 水麦冬科 Juncaginaceae

Triglochin maritima 海韭菜/37
Triglochin palustris 水麦冬/38

51 泽泻科 Alismataceae

Alisma orientale 东方泽泻/138

52 花蔺科 Butomaceae

Butomus umbellatus 花蔺/227

53 禾本科 Poaceae (Gramineae)

Agrostis gigantea 巨序剪股颖/39
Anthoxanthum glabrum 光稃香草
(Hierochloe glabra 光稃茅香)/40
Aristida adscensionis 三芒草/41

Aristida grandiglumis 大颖三芒草
(Stipagrostis grandiglumis 大颖针禾)/42
Aristida pennata 羽毛三芒草
(Stipagrostis pennata 羽毛针禾)/43
Arthraxon hispidus 荩草/44
Avena fatua 野燕麦/45
Bothriochloa ischaemum 白羊草/46
Bromus ciliatus 缘毛雀麦/47
Bromus ircutensis 沙地雀麦/48
Calamagrostis pseudophragmites 假苇拂子茅/49
Crypsis aculeata 隐花草(扎股草)/50
Cynodon dactylon 狗牙根/51
Dactylis glomerata 鸭茅/52
Digitaria ciliaris var. chrysoblephara
毛马唐/53
Elymus alashanicus 阿拉善披碱草
(Roegneria alashanica 阿拉善鹅观草)/54
Elymus dahuricus var. cylindricus
圆柱披碱草 (Elymus cylindricus)/55
Elymus excelsus 肥披碱草/56
Elymus tangutorum 麦宾草/57
Eriochloa villosa 野黍/58
Festuca dahurica 达乌里羊茅/59
Festuca rubra 紫羊茅/60
Helictotrichon schellianum 异燕麦/61
Melica scabrosa 臭草/62
Melica turczaninowiana 大臭草/63
Melica virgata 抱草/64
Phalaris arundinacea 虉草/65
Phragmites australis 芦苇/66
Poa pratensis ssp. angustifolia 细叶早熟禾
(Poa angustifolia)/67
Poa pseudopalustris 假沼早熟禾/68
Poa psilolepis 光稃早熟禾/69
Poa sphondylodes 硬质早熟禾/70
Ptilagrostis pelliotii 中亚细柄茅/71
Puccinellia distans 碱茅/72
Setaria glauca 金色狗尾草/73
Setaria viridis 狗尾草/74
Stipa aliena 异针茅/75
Stipa bungeana 长芒草(本氏针茅)/76

54 莎草科 Cyperaceae

Carex leiorhyncha 尖嘴苔草/346
Carex pediformis 脚苔草(日荫菅、柄苔草)/347
Carex stenophylloides 砾苔草/348
Kobresia humilis 矮生嵩草/349
Scirpus planiculmis 扁秆藨草/350

55 百合科 Liliaceae

Allium altaicum 阿尔泰葱/137
Allium sacculiferum 朝鲜薤/228
Allium stenodon 雾灵韭/340
Allium victorialis 茖葱/138
Asparagus schoberioides 龙须菜/77
Asparagus trichophyllus 曲枝天门冬/78
Gagea pauciflora 少花顶冰花/210
Hemerocallis lilioasphodelus 北黄花菜/211
Lilium dauricum 毛百合/237
Lilium martagon var. pilosiusculum
新疆野百合/269
Polygonatum sibiricum 黄精/139
Veratrum nigrum 藜芦/345

56 鸢尾科 Iridaceae

Iris loczyi 天山鸢尾/341